MOSE

THE MOSE EFFECT
THE CHALLENGES OF A PROJECT FOR THE FUTURE

NACANCO
NOLEGGIO

CT17

CT15

↑ The barrier at the Chioggia entrance during test operations• The Lido lagoon entrance. In the center, the new island linking the two barriers at Lido San Nicolò and Lido Treporti • The Malamocco entrance with the barrier during operations • The Chioggia entrance. On the left, the seaward sheltered harbor with the caissons beforethey were sunk • The floodgates in the storage area ready to be transported and installed. The male elements of the hinges have already been assembled • Detail of a caisson with compartment for the floodgate, showing the female elements of the hinge, before launch and installation • The jack-up during the installation of a floodgate for the Lido San Nicolò barrier.

Iconographic research and captions by
Elena Zambardi, Consorzio Venezia Nuova

Iconographic archive
Ministero delle Infrastrutture e della mobilità sostenibili – Provveditorato Interregionale per le opere Pubbliche del Triveneto
Concessionario Consorzio Venezia Nuova

Photographs
Archivio di Stato di Venezia, archivio fotografico
Bibliothèque Nationale de France
Massimo Belluzzo
Francesco Bonivento Lux Photo & Video
Marta Buso
Alessandra Chemollo
Consorzio Venezia Nuova
Emanuele Coppola CEMultimedia
DroneXplore
Fotorenard
London, © Trustees of the British Museum,
Giorgio Marcoaldi CEMultimedia
Andrea Merola
Rijkswaterstaat / Rob Doolaard, Bart Van Eyck,
Joop Van Houdt, Harry Van Reeke
Studio IKO

Design and layout
Tapiro

Translation from Italian
Lucian Comoy
for Language Consulting Congressi - Milan

First edition July 2021
ISBN 9788829713349

www.marsilioeditori.it

Available through ARTBOOK | D.A.P.
75 Broad Street, Suite 630
New York, NY 10004
www.artbook.com

Daniele Franco, *Minister of Economy and Finance*

"Let me reiterate: Water equals time and provides beauty with its double. Part water, we serve beauty in the same fashion. By rubbing water, this city improves time's looks, beautifies the future. That's what the role of this city in the universe is."[1]

With these words, Joseph Brodsky, poet and Nobel Prize laureate in literature in 1987, identifies the distinctive feature of Venice's beauty: the inextricable bond between land, water, and flowing time.

Of the surface area of the lagoon, a mere five per cent is made up of dry land, of islands. It is on this tiny percentage that Venice stands, an urban network crossed and surrounded by water.

Its relationship with water is the essence of its identity, but at the same time the source of the threat that has always made it naturally vulnerable.

The threat of the latest century is the acqua alta, *or high water.*

The acqua alta *phenomenon began to afflict the lagoon since 589* C.E.*; it then intensified since 1867 and reached truly perilous levels towards the end of the last century and the beginning of this one. It is the result of the simultaneous lowering of land (subsidence) and the rising of the sea (eustasy), which are in turn caused by natural climatic factors and anthropogenic causes, namely, by human intervention. It is also influenced by the emission of greenhouse gases, which contribute to an increase in global temperature and a rise in tidal levels.*

The MOSE (Modulo Sperimentale Elettromeccanico, Experimental electromechanical module) is the hydraulic system built to protect Venice from particularly high tides. It consists of a system of retractable dikes that are activated at the three lagoon inlets when the tide exceeds a certain threshold. The key word in the description of MOSE is retractable. When the tide goes out, its movable structure allows the caissons to be flooded and dropped back into their hinged casings on the sea bottom, so as not to hinder the exchange of water and passage of vessels, and so as not to be visible. This distinguishes MOSE from the imposing structures of dams built with the same function around the world at river mouths, such as the Thames Barrier or those of the Scheldt.

The work, originally conceived in the 1980s and then started in 2003, is now operational. It has been a complicated journey, not without heated discussions and critical moments.

MOSE thus takes up the baton in the series of extraordinary engineering works built to protect the city.

Since the foundation of Venice, every era has applied its technology to the creation and protection of a particular urban model, based on the balance between the city and the lagoon.

We may recall the piles used to construct buildings, with millions of oak, elm, and alder trunks pointed and driven down into the lagoon's clay to support the structures. Or the massive hydraulic works carried out to divert

the Bacchiglione, Brenta, and Piave rivers from the lagoon between the fourteenth and nineteenth century. These latter works defended the city from the other natural hazard that has historically threatened its survival: the silting up of the lagoon.
Humanity's determination to intervene in nature has benefited from ingenuity, creativity, and technological advancement and has in turn enhanced them. Italian technology is strongly characterized by its ability to adapt to change and react to challenges.
While MOSE will protect the lagoon for decades to come, it is important to remember that large parts of many regions of the world, in coastal areas or close to river deltas, may be affected by rising sea levels.
We can no longer postpone tackling the global climate emergency. A decisive, concrete, and coordinated response is needed. No country can go it alone, and no country should be exempt from contributing to the reduction of greenhouse gas emissions.
The Italian Presidency has put climate and environmental issues back at the heart of the G20 agenda. We are working, for example, to define common global standards for the dissemination of reliable and qualified information on the climate impact of production activities. Secondly, we are promoting a systematic dialogue on environmental taxation, in particular on carbon pricing policies.
Our focus on climate is reflected in the decision to hold the meeting of G20 Ministers of Finance and Central Bank Governors and the Climate Conference in Venice.
Venice was chosen not only for its artistic, historical, and cultural standing, but also as a symbol of our fragility and, at the same time, of our great capacity to adapt and react to adversity, of our tenacity and determination to safeguard vulnerable and precious natural habitats.

1. Joseph Brodsky, *Watermark* (New York: Farrar, Strauss & Giroux, 1992). Brodsky, who was born and lived in Russia until the age of thirty-two and then became a naturalized American, is buried in Venice, on the island of San Michele. Deeply attached to the lagoon city, Brodsky wrote his essay *Watermark* in 1989 at the invitation of the Consorzio Venezia Nuova.

Luigi Brugnaro, *Mayor of Venice*

The date of October 3, 2020 will forever be remembered as the day Venice stopped the sea. Throughout its 1,700-year history, the city has been a place of progress, cultural exchange, discoveries, ambassadorships, trade, integration, and freedom, but above all it has been a place where humankind, through its ingenuity and farsightedness, has employed science and intellect to tackle difficulties. And, first and foremost, the great challenge of Venezia Serenissima *has always been to "dominate" the waters.*
Our tradition reminds us of this every year with the Festa della Sensa *when, through a symbolic ring thrown into the waters of the lagoon, Venice marries the sea "as a sign of perpetual dominion." These ancient gestures are linked to events in our history but, like all Venetian traditions, they serve to remind us of who we are, what our ancestors succeeded in building and, above all, of the responsibility to continue along the path that makes Venice the oldest city of the future.*
Over the centuries we have been taught this: how to rule the seas, divert the rivers, clean the canals, safeguard the lagoon and its delicate ecosystem. And this is exactly why we have a duty to continue, with total respect for the environment and nature.
MOSE fully embodies this spirit. I have always been a firm supporter of this project because it represents a concrete response to the devastating effects of climate change on our city. This is why, today, we must ask ourselves how we intend to protect the environment.
We have before us two different approaches to tackling an issue concerning the future not only of our city but of the whole world and above all of those who live in it. On the one hand there are those who I would venture to call "environmentalist deniers," those who think that in order to safeguard our territory it is necessary to let nature take its course in a sort of uncontrolled self-regulation that humanity must endure. On the other, there are those who believe in a more "sustainable" vision in which our ingenuity and scientific discoveries must be made available to preserve resources for future generations, but above all to mitigate or perhaps prevent the destructive force of nature from generating situations of danger for humanity itself.
Our city has made the clear choice of believing in technology and engineering to address the issues of major climate change, which do not only refer to aqua granda, *or the tidal flooding of the city, but also rising sea levels and the repercussions this may have on islands. It is a gesture of responsibility towards all those children who have no vote as yet.*
I still have a vivid memory of every moment of that terrible aqua granda *of November 12, 2019 and all the disastrous consequences on the city. We could no longer remain passive spectators of decisions taken elsewhere. We could no longer ignore what fate the MOSE had, but above all we no longer wanted to be excluded from the events that concerned us. The* "Comitatone," *or Large committee, of November 26, 2019 marked the beginning of the turning*

point. I requested the establishment of a control room that would meet in the Prefecture and become a permanent fixture to update the city on the work being done to complete MOSE. The first meeting was called on January 7, 2020 and we sent out a strong and clear message: finish the work within a well-defined deadline to prevent further exceptional floods from once again bringing Venice to its knees.

And so it has been. I must thank Commissioner Elisabetta Spitz, who has shown tenacity and great cooperation with the city along the way. But along with her, I would also like to applaud all those who, each in their own area of competence, have rolled up their sleeves to complete the entire project. The embankment works at the Lido, those at Porto Marghera, and many other complementary works to protect the lagoon are yet to be completed, but we must remain confident. We want to repeat once more that Venice wants its own voice back. When the project is definitively handed over to the state and made operational, the Municipality and the Metropolitan City must play an active role in its management.

With the raising of the floodgates on October 3, 2020, and with all those that have followed, we have shown that we know how to work as a team and that the city will always be ready to play its part.

Venice and all its inhabitants have seen for themselves the value of a project that until that day had been, despite itself, a symbol of malfeasance and corruption, but which has finally begun to be talked about for what it really is: an indisputably essential infrastructure that is the result of our country's excellence in the field of engineering. After its detractors' many protests, demonstrations, and toxic statements, we have finally proved what we all expected: MOSE has won the challenge with the tidal waters and preserved the city.

Our commitment, therefore, must be not to minimize the environmental risks we may have yet to face but to enhance the technological systems to overcome them. The first true form of prevention is not being afraid to believe in progress. Let us team up and leave a tangible sign of our work. Let us do it with a great deal of foresight and a free spirit. Let us do it so that future generations can appreciate that, thanks to our commitment and courage, we have contributed to saving Venice, the world's heritage.

MOSE and Venice: the courage to invent

Elisabetta Spitz

A major public work is very much like a monument. It has a similar grandeur, cost, and claim to durability.
But it is not a monument.
If anything, a public work today is a crossroads. A crowded place, full of the projects of those who conceived it, the expectations of those who are destined to use it, and even the prognostications and forecasts of those who look further ahead. A junction between the past and the future. Something alive, in short, that protects the people it is intended for and helps to ensure that their lives are in some way enhanced.
We can also say that these works are often the stimulus for a broader public discourse, an opportunity to better focus on our objectives, our strategies, and the profile of our societies. In a word, a public work, any public work, is at one with the context in which it is placed. At the same time, it reveals this context and it moves it.
This applies all the more so to MOSE, the technological barrier whose function is to defend Venice from *acqua alta,*or tidal flooding, and which is a full-fledged part of that city, itself so marvelous and yet so special. It is part of it, living its life, protecting its public health, contributing in some way to its beauty and, finally, offering itself as its most outlying but decisive periphery. A protective and strategic periphery.
The history of MOSE, as we all know, began in the 1980s. It is the result of a unique, innovative, and wholly Italian project that cannot be compared to any of the many hydraulic defense projects that have been built in other parts of the world.
The need is not ours alone. Since ancient times, humanity has had to try to protect itself and the places where

← Detail of a barrier in operation.

progressive settlement took place from the fury of the sea and the elements. As a result, engineering has gradually become not only a tool for technological and infrastructural development, but also a peaceful weapon for the defense of man and his environment.
Clearly, this need has grown over time. And, as is all too easily foreseeable, it is destined to require ever greater commitment as "progress" forces us to come to terms with the environmental imbalances that all too often accompany its development.
There is therefore something universal in this story of ours. But also something very, very singular. For all dikes, dams, and defense structures against the aggression of the sea (and sometimes, more rarely, of large rivers) have been designed and built as large horizontal barriers to protect low-lying land or areas at risk of flooding. The large barriers built in the Netherlands or at the mouth of the Thames function as massive breakwaters that are moved like huge sliding panels when they have to allow boats to pass. MOSE, on the other hand, is a vertical barrier, so to speak, which rises from the sea only in emergency conditions, when it is called upon to protect the lagoon area from the risks posed by exceptionally high tides. The concept is different, the technology is different, and the impact on the environment is different.
The complexity of this mechanism is based on the simplest of principles, that of Archimedes: "The upward buoyant force that is exerted on a body immersed in a fluid is equal to the weight of the fluid that the body displaces."
Thus, based on this principle, imposing structures the size of buildings up to ten stories high (just to give you an idea of the size), placed and hinged on containers fixed to the seabed, rise when they are filled with air and become lighter. At this point the seventy-eight MOSE gates, divided into four barriers, suddenly emerge from their underwater enclosure and—being mounted side by side and perfectly aligned—stand in defense of the lagoon, preventing the sea from invading when the tide level is particularly high and dangerous.
If I may add a personal note, I would like to say that watching the rise of these barriers side by side at close quarters arouses an indescribable emotion. It is as if a number of giants were rising blithely from the sea, forming in unison, each bound to the other, an impassable wall capable of protecting a fragile and suggestive city that is largely defenseless against the not always benevolent force of nature. As soon as the barrier is erected and completed, it is possible to see in real time how

the sea is stopped and how, in the space of just a few minutes, the difference in level between the water outside and inside the barrier becomes clear. At high tide, the difference between the sea and the lagoon can be as much as 70 or 80 centimeters, and this is clearly visible on the two sides. On one side there is a rough or even stormy sea. On the other is a calm, flat, peaceful sea. On one side, there is the force of nature. On the other, the human ambition to keep it under control as far as possible.

This is the impressiveness (and appeal) of MOSE. But then there is its fragility, the risks it may run, the erosion of time that threatens it. Like all bodies immersed in salt water, MOSE, too, is subject to the aggression of the sea and brackish water, and it must be acknowledged that all the studies carried out in recent years on how to prevent corrosion of the metal components of underwater barriers have not produced solutions that we can consider definitive.

At present, the "useful" life of the barriers is estimated to be a maximum of one hundred years, provided that ordinary maintenance is carried out every five years and comprehensive maintenance every ten. The maintenance cycle of all submerged components requires a level of timeliness and precision that has not always been met.

The most delicate moment for the barriers is perhaps the one we are going through now, since the first provisional maintenance cycle has not yet begun, even though some barriers were assembled and installed almost seven years ago. In short, there is a delay to be made up for, and from then on a schedule to be met with absolute, strict punctuality.

As is well known, the MOSE construction site should be completed by the end of 2021. For the time being, 95–97% of the work has been finished, with very few details of the systems yet to be finalized, after which the work can be said to be complete and the construction sites (started back in 2003) can finally be closed down. But from then on, it is maintenance that will make the difference between the full success of the project and the risk of running into difficulties. It is at that point that the real start-up of MOSE will begin and with it the first cycle of provisional maintenance. To cope with this phase, two extra gates have already been built for each barrier so that it will be possible to replace two "pieces" at a time by means of a purpose-built vessel able to lift and immerse a gate and enable it to be attached to the hinge that keeps it in the stand-by position on the seabed. Replacing two gates at a time and bringing them to a construction

site for comprehensive maintenance will enable the entire maintenance cycle of the gates to be completed in ten to twelve non-consecutive months, at a rate of two gates per barrier per month.
It will therefore take at least two years to complete this provisional maintenance work on the gates, and only at that point will it be possible to define with absolute certainty the frequency necessary to carry out full-scale maintenance. This is a sort of work in progress that will inevitably involve a much wider audience than the limited circle of those involved.
In the meantime, research on the materials used in the construction of MOSE will continue and we can reasonably hope that in the not too distant future maintenance operations that today are so complicated will become increasingly effective and, it is to be hoped, less frequent than at present.
In short, we need to think about the future and start thinking now about how this whole situation can and must evolve and change.
The research underway could contribute to an improvement and greater effectiveness in the methods of planned maintenance of the mechanical and electronic components of MOSE. But this is only a first step. The need to think about the next phases is already pressing.
The two aspects on which work will have to be carried out in the next few years to make MOSE more effective concern, on the one hand, the system for supplying power to the barriers and, on the other, the system for forecasting the tides so that they can be raised in a more targeted manner during major surge events.
Currently, MOSE has been raised twenty times during the 2020 winter season with the aim of protecting inhabited areas from tides exceeding 130 centimeters. But this first and as yet still experimental season aimed at containing exceptional tides has already made us aware of the fact that it may not always be necessary to raise all the barriers at the same time. The weather events we encountered during this period presented themselves in different ways. Waves, winds, tides, rainfall, inflows of fresh water from rivers—all these phenomena affect the lagoon, giving rise to a number of wholly original combinations that rarely repeat themselves in a completely uniform manner. We have learnt that each event is different and unique, and that we therefore need to refine a containment strategy able to deal flexibly with all these differences each time. The raising of barriers will increasingly

↑ Venice, high water on October 29, 2018.

have to be calibrated to the peculiarities of the events they are called upon to deal with.
The lifting of individual barriers must therefore be anticipated or delayed on a case-by-case basis. We need to try to formulate more reliable and targeted forecasts of the lifting required, avoiding false alarms and underestimates. All this will also contribute to more effective management of the two ports (Venice and Chioggia) whose activities we must safeguard, while also minimizing the effects of closures on the lagoon ecosystem.
In other words, the technological power of MOSE will need to be graduated, managed with care, and used to the right extent and no more. It will not be a blind force with automatic reactions, but a motivated one that the experience gained over time will calibrate each time as required.
The other aspect that will require resourcefulness and an innovative approach is the energy supply for the lifting system.
At present, in order to push compressed air through the gates and keep it under pressure, it is necessary to use 1 GWh/year to ensure the operation of the barriers and 14 GWh/year to maintain the barrier and equipment complex. This requires an enormous consumption of energy. Naturally, MOSE's power system was designed more than twenty years ago, when "clean" energy sources as an alternative to more traditional ones were not yet available.
Now that renewable energies are changing this scenario, it is reasonable to assume that the barriers can be powered by energy produced from various alternative sources combined with each other. This would have advantages in terms of both costs and pollution. In particular, with the support of ENI, we are already considering how to develop a project to power the barriers using renewable energy sources that will help us make that ecological and environmental leap that is now inexorably part of our public agenda.
Of course, all these perspectives will have to come to terms with the environmental changes looming on the horizon. And these will be major changes requiring mankind and society to rethink their nature and which will subject large-scale projects to a real stress test. This will be a crucial time for MOSE too.
For now, we have to deal with forecasts and prospects that make us think, make us hope, but also make us fear. A great deal.
Consider these figures alone: "In the last 200 years," and

↓The Malamocco barrier in operation during a test

I quote from a study by geomorphologist Fabrizio Antonioli of ENEA, "the average level of the oceans has risen at a faster rate than in the last 3,000 years, with an alarming acceleration of 3.4 mm per year in the last two decades alone. Without a drastic change in direction of greenhouse gas emissions, the expected rise in sea level by 2100 will irreversibly alter the current morphology of the Italian territory, with a forecast flooding of up to 5,500 km^2 of coastal plain, where more than half of the Italian population is concentrated."
The territory between Trieste, Venice, and Ravenna is indicated by ENEA in the mapping of Italian coasts as being an area at risk of flooding as early as 2050. According to these scenarios—which are not as fanciful as we would like to think—in 2100 the entire Po Valley could be flooded due to the progressive rise in level of the northern Adriatic Sea. By the same date, some experts warn of the risk of flooding in the Gulf of Taranto and the plains of Oristano and Cagliari. Almost the whole of coastal Italy could find itself at risk of progressive erosion of its territory: from Tuscany (Versilia) to Lazio (Fiumicino, Fondi, Agro Pontino) and on to Campania (Sele and Volturno plains) as well as in Sicily (coastal area of Catania, Aeolian islands).
It is clear that this scenario poses dramatic and worrying questions for our society, which will probably require a thorough rethinking of all our strategies. All the more reason to ensure that the tidal containment measures of which MOSE is a forerunner be implemented and made available to respond to the needs—and emergencies—that the future holds in store for us.
The protection of coastlines and islands increasingly exposed to the progressive aggression of the sea will be entrusted to many other barriers that will need to be designed on the MOSE model. And by taking the idea further, of course. We are talking about dikes and gates emerging on demand; Italian engineering has been working on these for some time and we expect even greater results from them in future.
In the meantime, MOSE protects the Venice lagoon up to a sea level rise of about 3 meters above the current level. We can say that at least it is a good start.

→ The Malamocco barrier in operation during a test
↑ Venice, the Arsenale seen from the north

Venice and its lagoon
a world laboratory

Pier Vellinga

Sovereignty and innovation

Venice today illustrates how this city has been one of the greatest cultural and commercial powers in European history. The city in the midst of a lagoon protected by tidal channels was accessible only to ships and traders that understood the dynamic interaction of sediment and sea. Local sovereignty provided opportunities for innovation in shipbuilding, finance, and book printing in the fifteenth century. Later, various coastal cities discovered and followed similar trajectories such as London and Amsterdam in the seventeenth century and New York and New Orleans in the nineteenth century. In Asia, similar developments occurred in the past but accelerated in the twentieth and twenty-first century as can been seen in Shanghai, Ho Chi Min City, Jakarta, Bangkok, and many other delta cities.

The sea is your friend and sometimes your enemy. Living on the coast implies permanent exposure to storm surges and flooding. Storms and wave action cause erosion in some places and sedimentation elsewhere. Cities have been lost to the sea such as Heraklion (Alexandria) on the Egyptian coast and Dunwich on the English North Sea coast. Dunwich was one of the largest cities in twelfth-century England. The town was largely lost to the sea by a succession of storms in the thirteenth and fourteenth centuries.

Sedimentation can be just as disastrous. Brugge in Belgium is an example. In the fourteenth century it was one of the most important commercial ports in northern Europe with commercial and cultural links with Venice. However, the tidal channel connecting Brugge to the sea (the "Zwin") filled with sediment and commercial trading declined rapidly. In fact, Venice faced a similar problem one century later. The Venice

← Salt marshes in the Venetian lagoon

lagoon was slowly but continuously being filled with sediment carried by the rivers debouching into the lagoon. It became increasingly difficult for the ships to reach the port of Venice. Fortunately, at that time the Serenissima, as the republic of Venice called itself, had the innovative and successful foresight and courage to divert the rivers and their sediment away from the lagoon directly into the Adriatic Sea.

Lagoon and city are one

The Venice ecosystem geologically evolved in a dynamic equilibrium between sediment supplied by the river Po and sea level rise. For about a thousand years, the Venetians and their water authorities have continuously taken measures to conserve the lagoon system. This way a unique but managed ecosystem has evolved, an amalgam of coastal dunes and beaches, salt marshes, mudflats and shallows, and brackish and freshwater flora and fauna. Together with the historical city of Venice, it has been a UNESCO World Heritage site since 1987.

Venice and its lagoon should be seen as one (eco)system. The lagoon and its dynamic tidal channels cutting through shoals and flats once provided safety against invaders. At that time and even more so today, the lagoon and its islands and shoals provide safety against the Adriatic Sea waves and storm surges. Many coastal cities are protected by shallow seas and wetland ecosystems. The historic cities in the north of the Netherlands were and are protected by the Wadden Sea in a similar way. Many cities on the US east coast are protected by barrier islands against Atlantic storms and hurricanes. A typical example is Frisco beach, North Carolina, just after hurricane Isabel hit the coast in 2003; other examples are New Orleans on a delta with extensive wetlands; and Ho Chi Minh City, protected by the mangrove reservation Can Gio. These shallow coastal ecosystems protect the urban areas against storm surges and direct wave attack. However, there is always the danger of a breach as can be seen in Figure 3. The danger of breaching has well been recognized by the Venice Water Authorities. The sides of the islands giving onto the sea are highly protected, as can be seen for the island of Pellestrina.

The fate of delta cities: subsidence and sea level rise

Deltas develop where rivers reach the sea and drop sediments collected upstream. As long as the supply of sediment keeps up with sea level rise the delta will expand. Throughout history such deltas have been hotspots of urban development. Examples are the Nile river delta (Alexandria), the Yangste river delta (Shanghai), the Rhine/Meuse/ Scheldt delta (The

→ Venice and its lagoon in the fifteenth century, ideally positioned for safety and sovereignty. Hugues de Comminelles, *Venetia*, from Claudius Ptolomaeus, *Cosmographia*, interpreter Jacobus Angelus, 1470, Paris, BNF, ms. Latino 4802, fol. 132.

VENETIA
132

← ↑ Venice and its lagoon, Unesco World Heritage Site since 1987.

↑ Barrier islands protecting major US east coast cities. Cape Hatteras, Cape Point, Diamon Shoals, Outer Banks, North Carolina, US.
→ Coastal wetlands protecting New Orleans.
→ Mangrove forest protecting Ho Chi Minh City.

Netherlands), the Mississippi delta (New Orleans), and the Po delta (Ravenna)—and there are many other deltas and delta cities throughout the world. The shallow coastal lands provide excellent opportunities for human settlement. But as time proceeds there is a price to pay: subsidence. Delta cities are sinking cities.

Deltas are rich in vegetation that flourishes and dies with every new season. The vegetation has accumulated and turned into thick layers of peat over the thousands of years before human settlement. These peat layers start shrinking as soon as the water is drained for urban development, industrial development, or agriculture.

As water drainage is a prerequisite for urban development, subsidence and sinking are inevitable consequences of human settlement in deltas. Water withdrawal for human consumption or industrial purposes usually aggravates the subsidence problems.

In Venice, the process of subsidence must have started when the first port facilities, warehouses, and palazzi were built about a thousand years ago. Subsidence must have been more or less in pace with the periodic renewal of the buildings and infrastructure. At that time, it was not a barrier for growth because the Venetians were well aware of sea level rise and subsidence. They built pile foundations with waterproof stones on top and made sure their floors were well above high water level. Venice could reasonably keep up with subsidence until the twentieth century, when massive quantities of water were withdrawn from the subsoil to support rapid industrial development.

The accelerating subsidence was hardly noticed until November 1966, when a major storm surge flooded the entire city. The city's sinking was identified as the major cause. The population in Venice, Italy, and worldwide was shocked. Something should be done, was the general political consensus. Both the historic city and the lagoon were at stake. The pumping of water from the subsoil was stopped, but that was not enough. The vulnerability of the city would continue to increase because of sea level rise. The 1966 flooding event eventually triggered a major safeguarding project with a crucial role for the innovative works we now know as MOSE.

Subsidence of cities and their protective wetlands

At the present moment, nearly all delta cities are sinking, including Shanghai, Jakarta, Bangkok, Manilla, and Ho Chi Minh City. One of the main reasons is drainage and water withdrawal. The Netherlands has a long history of subsidence

↑ → Island of Pellestrina and the Venice lagoon protected against storm waves by barrier islands.

and flooding as well as subsequent coastal protection works. More than a thousand years of water drainage, subsidence, and dike building is the main reason why most of the population now lives below sea level.

Fortunately, local decisions and local measures can stop this process considerably, as has been demonstrated by Venice and Tokyo. John Englander, an American oceanographer and sea level rise expert, made an inventory of sinking cities, including
a ranking: 1st Jakarta, 2nd Manila, 3rd Ho Chi Minh City, 4th New Orleans, 5th Bangkok, 6th Osaka, 7th Dhaka, 8th Shanghai, 9th Venice, 10th Alexandria (Egypt).[1]

In the last half century, some cities have improved due to tight restrictions on pumping groundwater, such as in Tokyo and Venice. Others, such as Jakarta and Bangkok, are sinking faster.

While the urban centers are subsiding, their surroundings, including the wetlands, are also sinking. Being aware of the protective role of the lagoon the Venetian Water Authorities have been monitoring the lagoon and its depth for many ages. The records reveal that the sinking over the last seventy years is greater than can be explained by subsidence and sea level rise.

The reason is as follows: The rivers flowing directly into the lagoon including their sediment input were diverted in the fifteenth and sixteenth century. This solved the problem of shoaling of the shipping channels, but caused a kind of (sediment) starving of the lagoon.

On top of this, additional interventions affected the tidal in- and outflow of water. Existing shipping channels were deepened, a major new shipping channel was created right across the lagoon to provide access to the industrial port Marghera, and the breakwaters/guiding walls at the entrances of the lagoon were extended into the Adriatic Sea.

Together, these interventions affected the sediment flows in such a way that the lagoon started to export sediment to the sea. The lagoon has deepened, and many small islands, shoals, and salt marshes have disappeared.

As a consequence, the shallow water system is slowly changing into an open sea system. Once this was recognized, various measures have been taken to protect the shoals and marshes, such as revegetation and protection by stones.

However, as long as the export of sediment continues

→ Relative sea level rise is the sum of subsidence and sea level rise. The blue line illustrates that sea level itself had risen by about 250 mm from 1870 to 2014. The black line illustrates that the city itself experienced a rise in sea level of about 450 mm over the same period. The difference is the subsidence of the land, some 200 mm over that period. As can be seen, the rate of subsidence was much higher in the period from 1950 to 1970, the period of extra water withdrawal. When this came to an end, there was some rebound until about 1990.
↓ The deepening of the lagoon and the disappearance of shoals and salt marshes. Salt marshes coverage: 25% > 8% of the lagoon surface in 70 years; sediment loss at sea ~ 700.000 m³/year

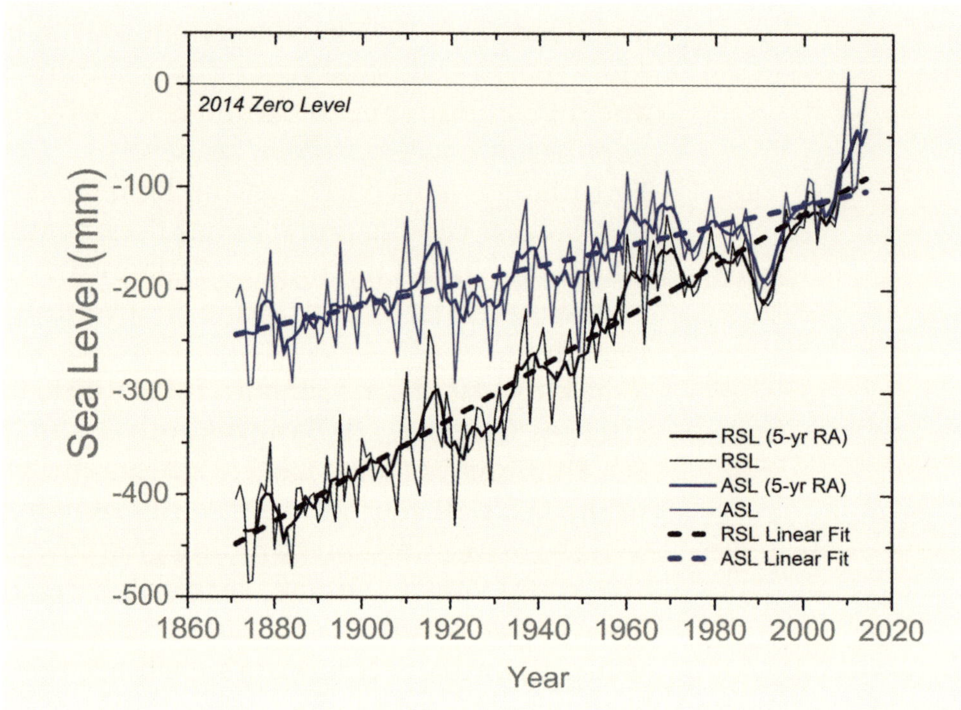

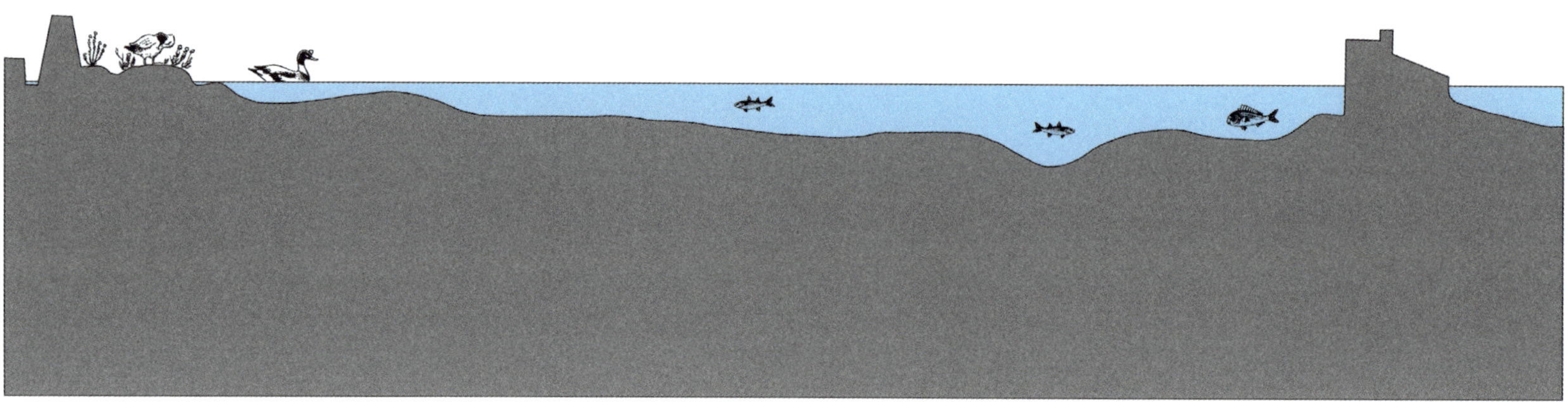

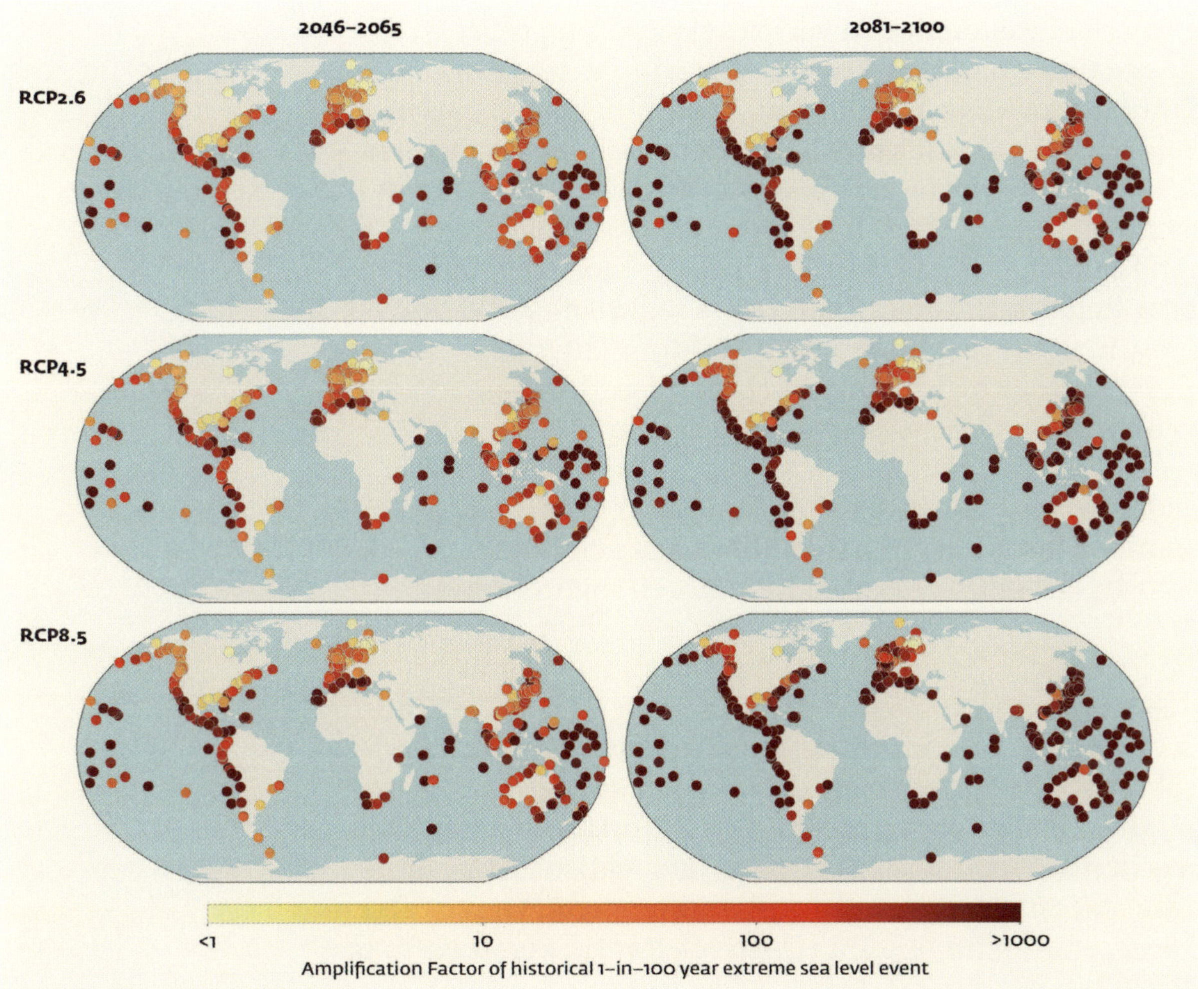

← The colors of dots express the factor by which the frequency of extreme sea level (ESL) events increases in the future for events which historically have a return period of 100 years. Source: IPCC, "Sea Level Rise and Implications for Low-Lying Islands, Coasts and Communities," chapter 4 of *Special Report on the Ocean and Cryosphere in a Changing Climate* (2019).

↓ Projected global average sea level rise (SLR) until 2300. Source: IPCC, "Sea Level Rise and Implications for Low-Lying Islands, Coasts and Communities," chapter 4 of *Special Report on the Ocean and Cryosphere in a Changing Climate* (2019).

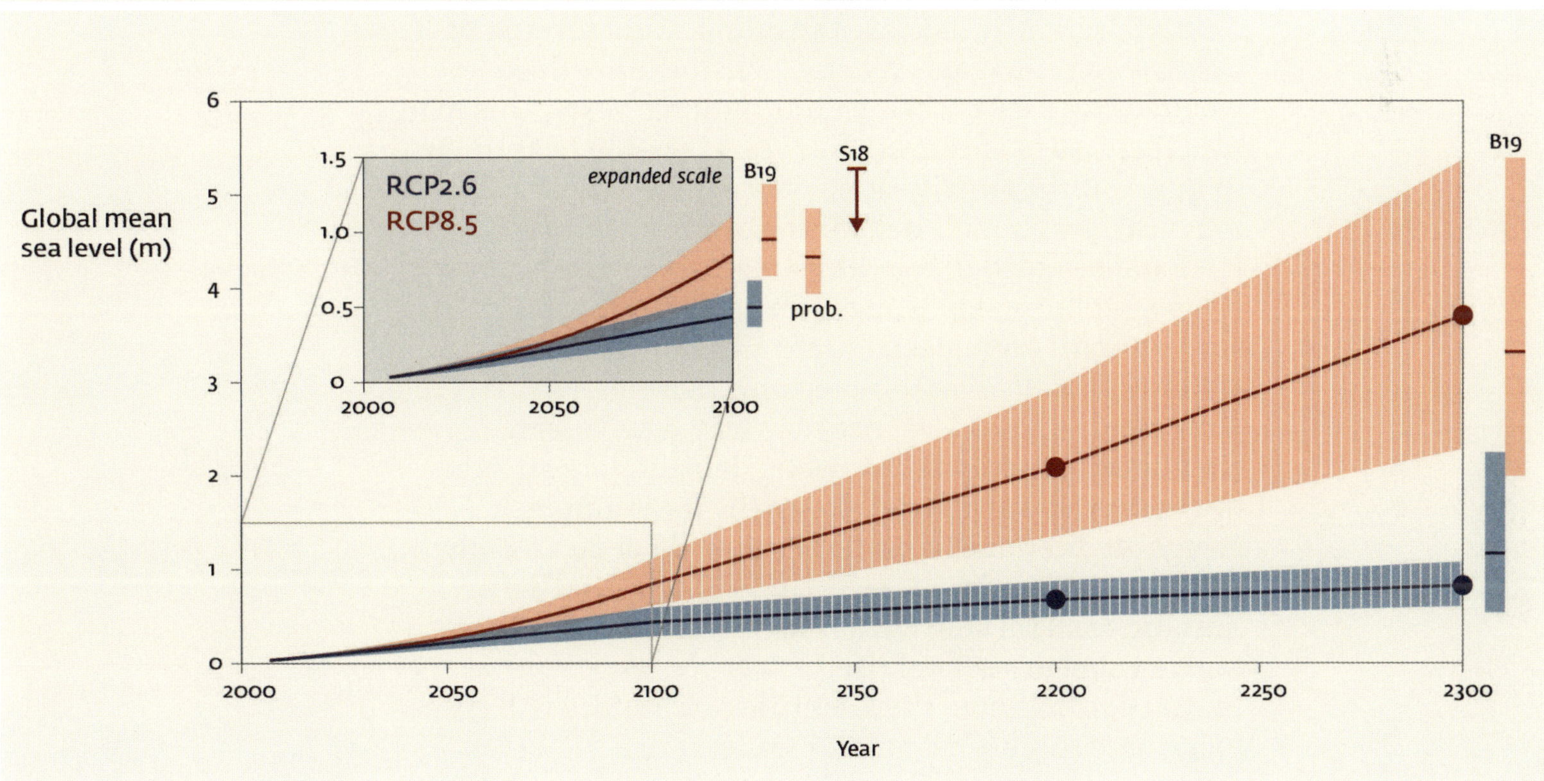

this will not be enough to reverse the deepening process. Venice again is an early example of dealing with delta dynamics, providing a full library of documented experience. The sinking of wetlands protecting delta cities is not unique to Venice. New Orleans and many other delta cities have the same experiences. Often there are additional processes at hand causing subsidence. One is the exploitation of oil and gas fields in the delta. Another is sediment starvation as a result of up-river (hydro)dam building. Living in a delta comes with opportunities but indeed there is a price to pay as these examples demonstrate.

Climate change and sea level rise

Subsidence can be controlled by local water management measures. This is not the case with climate change. Hopefully, the efforts to reduce climate change as agreed in Paris in 2015 will be successful. But even then, the melting of the ice caps of Greenland and Antarctica will not fully stop as seen in the projections presented by IPPC.[2] The RCP 8.5 scenario corresponds to a global average temperature rise of roughly 4 degrees centigrade while RCP 2.6 corresponds to roughly 1 degree centigrade. Given the ongoing worldwide efforts in reaching the Paris targets, it is realistic to assume that the outcome will be ranging between the RPC 2.6 and the RPC 8.5 scenario, implying a rise in sea level of about 1.2 m in 2200 and 2 m in 2300.
This will seriously increase the number and severity of floods over the coming years. Whereas, historically, flood levels would be reached once every one hundred years, this may well increase to once every five years over the next thirty year.[3] In figure 9, the amplification of the frequency of surpassing a critical level is indicated for the coming century.
It should be noted that local sea level rise is likely to vary by geographic region. Long term geological processes such as uplift after glacial times and changes in ice cap volumes as well as local geotectonic movements play a role. Moreover, local sea level rise may vary when ocean circulation patterns and seawater densities are affected by the overall warming of the earth. For example, a verage Mediterranean sea level rise will not necessarily equal global average sea level rise. In turn, Northern Adriatic sea level rise will not necessarily equal average Mediterranean sea level rise.
In the years to come, the scientific community will try to be more specific about local sea level rise. However, some uncertainty will remain.

Early signals, early plans

Worldwide observations of the last few decades illustrate an acceleration of the rate of global average sea level rise. In the period 1900 to 1990, the rate was about 1.5 mm per year. Between 1990 and 2020, the rate was about 3 mm per year.[4] A warmer earth and a warmer ocean will likely bring more intensive storm conditions and hurricanes.
While sea level is rising at the same time, such storms and hurricanes will be flooding a larger area. Hurricane Sandy, which hit New York in 2012, may be considered one of the first examples of climate change-induced floods. It was highly exceptional that a hurricane of such strength would reach New York.
Meanwhile, the city had become a little more vulnerable because of the extra sea level rise. The damage caused by hurricane Sandy was estimated at 70 billion dollars. New York city, including Manhattan and its metro system, was more vulnerable than expected. A range of protection schemes has been proposed by Jeroen C. J. H. Aerts, W. J. Wouter Botzen, Hans de Moel and Malcolm Bowman, published in the *Annals of the New York Academy of Sciences*.[5] The City of New York has recently started to upgrade the level of protection at the waterfront. Decisions about the building of one of more storm surge barriers were passed on to the future.
An additional example is Shanghai. This major delta city with a population of 27 million is suffering from subsidence. The frequency of flooding of the center area has increased significantly over the last few decades. The cumulative subsidence is about 2 m, mainly in the previous century. Since measures have been taken, the rate of subsidence has been reduced to almost zero, but the area suffering from subsidence has become larger.[6]
When evaluating measures to protect delta cities, a mix of techniques are usually explored[7]. Raising the streets and houses is an obvious one. Second is upgrading the natural protection systems such as coastal revetments and foreshores. Third is to build dikes and levees. Fourth is to build storm surge barriers. Retreat is usually not considered an option in the highly urbanized areas, and most often there is an economic rationale.
In evaluating protection strategies, usually a small number of concepts are applied such as cost benefit analysis, cost effectiveness, and nature conservation. However, sea level will keep on rising for many years to come even when greenhouse emissions are radically reduced.
Therefore, it is most important to ask what we will do next,

when sea level rises an additional 0.5 or 1 m. One of the most appropriate solutions for today may well be a barrier for implementing the next steps. While solutions for today can also be an effective first step in the long term. In scientific literature, the term "adaptive strategy planning" has been coined to describe the process of considering the long term in making decisions regarding the short term. As the long term is uncertain by definition, several strategy-trajectories can be explored in adaptive strategy-planning. As reality unfolds one can "switch" from one trajectory to another with minimized stranded assets.

Integral planning is another crucial element of successful adaptation. In most cases, the problems a highly urbanized area is facing are multifaceted. Subsidence, sea level rise, water drainage, drought and heat management, and biodiversity management are usually intertwined issues.

Venice and its lagoon, a world laboratory

Living in a delta is attractive and risky at the same time. Throughout history, Venice has been a pioneer in harvesting the benefits and finding new solutions. The building of MOSE is an awesome example. But not just MOSE. It is the combination of raising the quays and alleys, upgrading coastal protection, rehabilitating the lagoon ecosystem and the building of MOSE that is so valuable and effective in dealing with subsidence and sea level rise.

The city and its lagoon will be safe for the next fifty to a hundred years. MOSE is a robust structure to protect against sea level rise up to about 0.5 m. The frequency of closing will increase over the decades to come. This will allow the ecosystem to adjust step by step without major losses. When sea level rises beyond 0.5 m, additional measures need to be taken. Maybe the MOSE "doors" (now with a height of 12 to 20 m) can even be slightly lengthened. But, with more frequent closures, the tidal lagoon will gradually turn into a brackish coastal ecosystem. By then plans should be ready for the following steps.

The major merit of MOSE is the provision of time. Venice and the world are receiving an additional period of fifty to one hundred years.

What may happen over this period? In very speculative terms, we might find ways to reverse climate change and sea level rise by then. More realistically, however, by then all delta cities will have gone through their first phase of flooding events and their first phase of protection measures.

By then, like Venice, they will need to make really difficult

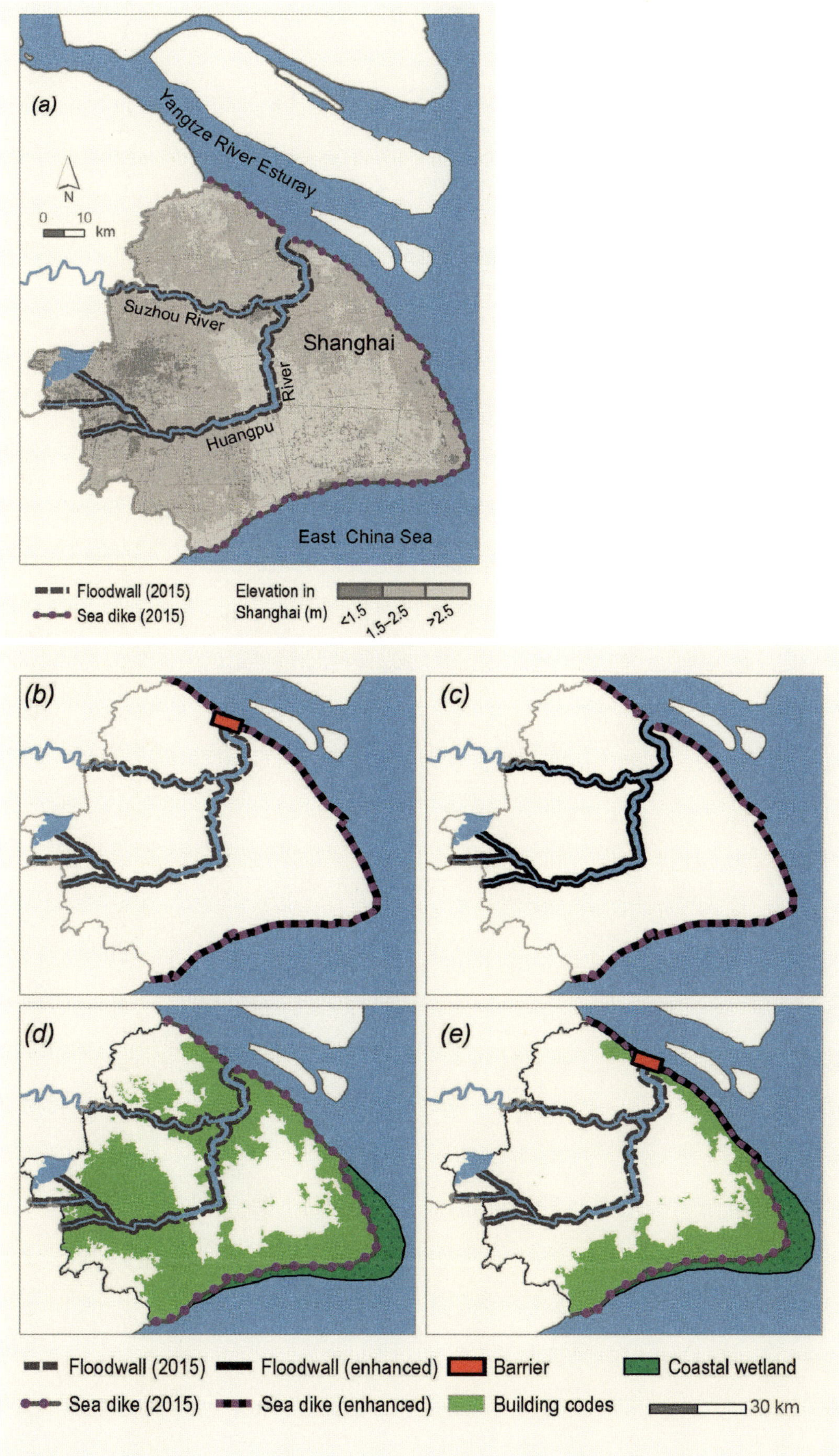

← Proposal of protective measures as a response to more frequent flooding, Shanghai.
a) 2015 situation
b) enhanced floodwalls and sea dikes
c) "soft" strategies that strengthen the building codes and natural coast
d) a hybrid strategy which combines elements from the other strategies
Source: Shiqiang Dua, Paolo Scussolini, Philip J. Ward, Min Zhang, Jiahong Wen, Luyang Wang, Elco Koks, Andres Diaz-Loaiza, Jun Gao, Qian Ke, and Jeroen C. J. H. Aerts, "Hard or soft flood adaptation? Advantages of a hybrid strategy for Shanghai," in *Global Environmental Change* 61 (March 2020)

choices on what to protect and where to retreat.
The population of Venice and the authorities in Venice and Rome have gone through a major experience in debating, designing, financing, and implementing the works now protecting Venice and its lagoon.
It has taken some fifty years to design, discuss, and reach agreement in particular about MOSE.
This is what it takes to develop and implement such complex plans for protecting delta cities. Venice is now ahead of its time.
The know-how and the experience gained is a precious example and a source of inspiration for all delta cities worldwide. If Venice can be saved, then all delta cities can be saved at least for the next fifty to one hundred years.

1. John Englander, "Top 10 Sinking Cities in the World," 7 January 2018, *johnenglander.net*, available at https://johnenglander.net/top-10-sinking-cities-in-the-world.
2. See Intergovernmental Panel on Climate Change, "Technical Summary", *Special Report on the Ocean and Cryosphere in a Changing Climate*, 2019, available at https://www.ipcc.ch/srocc/chapter/technical-summary/.
3. See Intergovernmental Panel on Climate Change, "See Level Rise and Implications for Low-Lying Islands, Coasts, and Communities," *Special Report on the Ocean and Cryosphere in a Changing Climate*, 2019, available at https://www.ipcc.ch/srocc/chapter/technical-summary/.
4. Thomas Frederikse et al., "The causes of sea-level rise since 1900," in *Nature* 584 (19 August 2000): 393–397; NASA Goddard Space Flight Center (PO.DAAC), https://climate.nasa.gov/vital-signs/sea-level/.
5. Jeroen C. J. H. Aerts, W. J. Wouter Botzen, Hans de Moel, and Malcolm Bowman, "Cost estimates for flood resilience and protection strategies," in *Annals of the New York Academy of Sciences*, August 2013.
6. Shiqiang Dua, Paolo Scussolini, Philip J. Ward, Min Zhang, Jiahong Wen, Luyang Wang, Elco Koks, Andres Diaz-Loaiza, Jun Gao, Qian Kec, Jeroen C. J. H. Aerts, "Hard or soft flood adaptation? Advantages of a hybrid strategy for Shanghai," in *Global Environmental Change* 61 (March 2020).
7. Lars T. de Ruig, Patrick L. Barnard, W. J. Wouter Botzen, Phyllis Grifman, Juliette Finzi Hart, Hans de Moel, Nick Sadrpour, Jeroen C. J. H. Aerts, "An economic evaluation of adaptation pathways in coastal mega cities: An illustration for Los Angeles," in *Science of the Total Environment* 678 (2019): 647–659.

Solutions and governance models similar to MOSE in Venice. Analysis of several international experiences

Anja Visini

Mobile barrier systems protect many vulnerable coastal locations from rising waters and some of these systems have been used as case studies for the MOSE project in Venice, although each context is physically different and tidal patterns and weather phenomena characterize each situation as unique.

In recent centuries, eustasy, which continues to increase due to the greenhouse effect, has led to a global rise in the average sea level. The need behind the construction of the barrages is, at least in part, the same as in Venice: to protect territory behind the defense from the phenomenon of subsidence and to counter the destructive effect of flooding. However, the conditions under which the works have been carried out, the surrounding constraints, the types of tide for which the barrages were planned, and the frequency of their operation are different in each case. Each system constitutes a site-specific project that establishes a compromise between environmental impact and the need to protect man-made areas, in which social, economic, natural and administrative factors are at play.

Since these are large-scale projects to manage highly changeable elements such as air and water—particularly when they act together—a certain level of uncertainty must be taken into account, especially at a time when dramatic climate change makes it difficult accurately to predict future situations. The scenarios produced by human activity are also a fundamental element that must be taken into account when planning. Therefore, research does not stop with the construction of the infrastructure but continues with an evolutionary process that sees the work, once built, as an

← Grevelingendam, the dike on the Rhine-Meuse delta (between the provinces of South Holland and Zeeland).

integral part of the territory, which in turn is influenced by it. The study of this type of system is a matter of ongoing research into the joint evolution of man and the environment, hopefully conducted with a view to collaboration rather than strenuous struggle and defense against nature.
In an ever-changing world, human infrastructure is not thought of as a series of eternal, unchanging works, but rather as systems to be assiduously monitored and maintained. Infrastructural design is therefore not limited to a mere engineering operation but becomes a field of environmental research on the one hand and management experimentation on the other.
One of the best-known and most impressive international experiences is that of the Netherlands, a system on a country-wide scale consisting of multiple projects to protect against high tides and sea storms, demonstrating how it is possible to plan for the very long term and at the same time operate effectively in the present.
The Delta Plan, consisting of thirteen projects, was drawn up in 1953 following the disastrous flooding in which more than 1800 people died and about 150,000 hectares of land were submerged, causing enormous economic damage.
The Dutch territory covers an area of 41,543 km^2, 40% of which below average sea level and 65% subject to storm surge levels. A purely passive defense policy, which began around the ninth century with the construction of sandbanks to protect the polders, was replaced in the fourteenth century by a phase of conquest of the areas of the inland lakes with the use of the famous windmills, and at the beginning of the industrial era with the land beginning to be reclaimed from the waters with the use of water-scooping machines, gaining over 5,000 km^2 of land located up to 7 m below sea level. However, the embankments did not guarantee safety in the face of extreme weather events, so between 1927 and 1932 the Zuiderzeewerken project was carried out, comprising a 32 km earthen dike closing off the inland sea of the Zuiderzee, together with the construction of 100 km of inland locks for the development of further polders. It was only after 1953 that the Delta Plan, completed in 1997, was conceived as a safe and definitive answer to the water management problem.
The Delta Plan chain of projects is located at the mouths of the great rivers Rhine, Meuse and Scheldt, which flow into the North Sea after crossing the Netherlands. In addition to protection from extreme events, a hydraulic system of river branches and barriers makes the entire area navigable,

↑ Brouwersdam, the dike between Goeree-Overflakkee (South Holland) and Schouwen-Duiveland (Zeeland).
← Bathse Spuisluis, the sluice gate situated near Rilland (Zeeland).

with the aim of living with the water and finding a compromise between man and nature. The works include the Eastern Scheldt Barrier (Oosterscheldekering, 1976–1986), the Maeslant Barrier (Maeslantkering, 1991–1997), the Hollandsche IJssel Barrier (1957), the Hartel Barrier (Hartelkering, 1991–1997), the Haringvliet Dam (1958–1971), Volkeraksluizen (1957–1969), Bathse Spuisluis (1980–1987), Brouwersdam (1962–1971), Grevelingendam (1958–1965), Veerse Gatdam (1958–1961, Zandkreekdam (1959–1960), Philipsdam (1976–1987), and the Ramspol barrier (2002). Taken together, these infrastructures constitute a true sample of large-scale engineering experiments.

One of the most famous and important installations is the Oosterscheldekering of the Eastern Scheldt, built to protect the city of Amsterdam; it is a 9 km-long antiflood barrier made up of pillars and sliding gates 6 to 12 m high, which move up and down and are closed when the water is expected to rise above the normal level of Amsterdam NAP (Normaal Amsterdams Peil).

The barrier provides not only a water containment device, but also a road link to the islands of Schouwen-Duiveland and Noord-Beveland in the province of Zeeland and a lock that allows fishing boats to reach the open sea. Included in the Eastern Scheldt national park, it provides for the protection of the marine flora and fauna thanks to the possibility of opening and closing the barrier.

The Maeslant barrier, on the other hand, is based on a rotating system: a mobile dam consisting of two 22 m-high, 240 m-long gates, which pivot around a metal sphere 10 m in diameter and weighing 680 t, which acts as a hinge.

Positioned on dry land, it allows both movement on the flat and upwards. Operated by an automatic system that is activated when the water level rises, the barriers, once closed, are filled with water so that the increase in weight causes them to rest on the submerged concrete foundations.

The Hollandsche IJssel stands between Rotterdam and the North Sea. Built to prevent stormwater from entering one of the most densely populated areas of the Netherlands, it is actually a dam on the river, as a result of which the barrier system would have led to overflowing of the watercourse, which would no longer have been able to flow into the sea.

To obviate this, two 80 m-wide locks were installed, suspended between towers, which are moved vertically if there is a risk of flooding, and flanked by a lateral channel which allows for maritime traffic.

← Hartelkering, the lock near Spijkenisse (Zuid-Holland).

The Hartel barrier is also constructed in a similar way, with oval towers from which elliptical containment systems are hung and then lowered into the water against the most impetuous storm surges. When they are completely submerged, they entirely close off the channel, with no possibility of marine passage, and for this reason the structure includes control portals on each abutment as well as in the middle, managed by a computer system.

Unique of its kind, the Ramspol barrier is designed with an inflatable mechanism, consisting of three rubber fabric gussets. Approximately 80 m long and 13 m wide, with a height of 5–8 m, they are the tallest inflatable devices ever made. The final shape is achieved by injecting air while a gravity feed system allows water to enter; by emptying the water and releasing the air, the barrier lowers again below the tide level.

Advanced technology and innovative character typify the remarkable structures of the Delta Plan, but the system is completed by a series of locks and fixed weirs such as the Haringvliet dam which, completed in 1971, serves to protect against flooding but also to drain the water of the Rhine and Meuse rivers.

It is an open structure with seventeen gates, which regulate the amount of water flowing into the North Sea, provide a lock for ships and special tunnels through which fish can pass. The condition for the fauna is one of the discriminating factors in the design of Dutch locks.

Fixed dams are often integrated with communication infrastructure such as roads and bridges: they block seawater, but at the same time regulate the inflow of fresh water into the system and control its backflow, as is the case with Bathse Spuisluis, Brouwersdam, Grevelingendam, and Veerse Gatdam.

Safeguarding the environment is fundamental to the operation of the plan, so the hydrogeological defense committees have a say in decisions on interventions and dialogue with the commissioner responsible to the Ministry of Infrastructure.

This does not detract from the fact that the Delta Plan has brought about considerable changes in the ecosystems of the areas involved, which have been the subject of specific environmental planning, and that the need to preserve the natural environment has for its part required substantial modifications to the initial proposals.

The need for open systems was discussed at the end

→ Philipsdam and Krammersluizen, the dike and locks located between Grevelingendam and Sint Philipsland (Zeeland).

← Volkeraksluizen, the locks between Goeree-Overflakkee (South Holland) and the province of North Brabant.
↑ Haringvliet dam at Hellevoetsluis, between Voorne-Putten and Goeree-Overflakkee (Zuid-Holland).

of the 1960s, when the implementation phase was already in full swing, due to a drastic reduction in water exchange in the basins, a decrease in saline concentration and an increase in pollutants from the rivers. The ecosystem of the estuaries would have been irreparably affected, as would the local economy, based on fish farming and tourism. In an ecological census, all plant and animal species were catalogued, and water samples were taken and analyzed to parameterize the entire wetland; the freshwater reserves were also protected from contamination by river pollutants and saline upwelling.
The environmental impact studies led to a structural change in the weir system and a redesign of the remaining works, incorporating infrastructure with sluice gates.
The relationship with water is not seen as one of contrast but as one with a precious element on which is based the economy of a delicate and highly productive area in terms of agriculture and energy.
Other works of considerable interest carried out in an international context are, on the other hand, individual infrastructures, not part of a system as is the Dutch plan.
Among the most significant are the Saint Petersburg dam, the Thames barriers, and the Nagaragawa barrier in Japan.
Saint Petersburg is a vital port and one of the most beautiful places in Russia.
The city lies on a coastal plain at just over 2.5 m above sea level, where the Neva flows into the Gulf of Finland.
Floods, including one in 1824 with levels more than 4 m above normal levels, have always been part of its history. The Saint Petersburg dam is able to protect the city from storm surges with waters up to 5 m high. Begun in 1978, it is a 25 km-long structure, consisting of a series of eleven separate dams; two openings allow ships to pass through and can be closed by six pivoting metal gates.
The entire complex also includes thirty water purification plants and transport infrastructure.
The Thames Barrier Scheme is smaller in size but highly technological, consisting of a 520 m structure with concrete piers, two abutment buildings, and a series of rotating steel gates.
Normally submerged and empty inside, these are filled with water and become operational in just 15 minutes, turning on themselves until the gates close. The system, built between 1974 and 1982, was supplemented by a plan to rebuild the embankments; over time the gates were activated from two to as many as twenty-eight times a year between 2013 and 2014.

← ↑ The Eastern Scheldt Barrier (Oosterscheldekering) located between Schouwen-Duiveland and Noord-Beveland (Zeeland).
↓ The Hollandsche IJssel barrier near Krimpen aan den IJssel (South Holland).

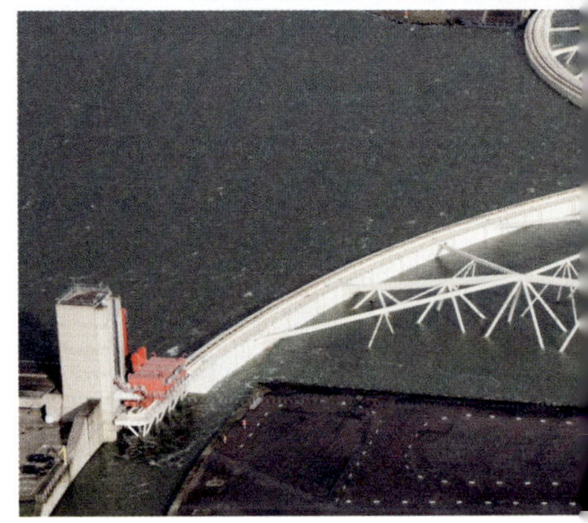

← ↑ The Maeslant barrier at Hoek van Holland at the mouth of the Nieuwe Waterweg (Rotterdam).

The mobile locks of the Nagaragawa Barrier in the Nagara River estuary, completed in 1995, use double barriers with a shell structure, which move vertically suspended between pylons.
This configuration allows overflow, the passage of river water over the gates, as well as underflow beneath them.
The barriers are configured according to the condition to be faced, the resulting inflow over the weir and the river level downstream, also taking into account possible seismic waves.
The examples listed above are case studies both in terms of technologies employed and as models of governance applied positively to complex systems involving diverse, complex, fragile territories.
The Dutch experts underline the importance and at the same time the difficulty of maintaining qualified teams, accustomed to the complexity of a high-tech system inserted in vulnerable areas.
In the case of the Delta Plan project, for example, there are several objectives that are checked annually, including flood risk, freshwater management, spatial adaptation of the works, as well as special decisions for certain regions such as the IJsselmeer and the Rhine-Meuse delta.
In comparison, the works on the Thames and in Saint Petersburg, although of great infrastructural magnitude, are simpler in terms of management and the number of problems in the area, and are not comparable to the interconnectedness of elements and critical aspects in the Venetian lagoon area.
The MOSE system and its general plan of interventions must take into account not only the increase in high waters, whose average annual frequency has increased from about ten in the first half of the twentieth century to almost sixty in the last five years, with an average of 80 cm more in terms of level, but also coastal erosion and morphological and environmental degradation.
The general plan of action combines the protection of the urbanized territory with the protection of the ecosystem, preparing distinct but interconnected lines of action.
The management can have a risk-based approach, simply establishing a safety level around which to set the defense of the areas concerned, or adaptive management, which also considers the uncertainties

↑ The Veerse Dam, the dike located between Walcheren and Noord-Beveland (Zeeland).
← Zandkreekdam, the dike located between Noord and Zuid-Beveland (Zealand).

in climate change and socioeconomic developments
to prepare best for future scenarios with flexible responses.
In order to deal with uncertain situations, it is necessary
to provide for the care and maintenance of the works,
assuring their safety objectives, but also to draw up
management plans for the lagoon and the coastline,
in order to mitigate the level of degradation and improve
the environmental and morphological qualities.
The protection of the natural and artificial environment
also means enhancing the value of this great work
of engineering, so that it can be linked to international
experience in the lagoon and maritime sphere,
with a view to cooperation and as an engine
for development not only locally, in the Venice
and Veneto areas, but also internationally,
as part of everyone's heritage.

Venice and innovation: water, land, and transformations in history

Elena Svalduz

Lagoon and city in motion

The lagoon landscape is subject to continuous transformation. A slow landscape, "powerful and at the same time reassuring" in its aesthetic fruition, "daily in movement in search of an ever-changing balance": in his "narrative report" conducted among stones, waters, and people, Francesco Erbani describes the ("mistreated") lagoon of Venice in this way.[1] Since its origins, it has been subject to changes caused by an overlay of factors and effects that have always threatened its survival: tributary rivers, erosion, bathymetric fluctuations, moon phases, silting, swamping... In emphasizing its uniqueness as an extraordinary heritage of environmental values, narrative and at times even contemporary historiography seem to have mostly privileged the natural component over the artificial one,[2] to the point of underestimating a fundamental aspect: that it is a context shaped by manmade action. From this point of view, the Venetian lagoon does not appear to be so far removed from the phenomena alluded to by Guido Piovene within the broader dimension of the Veneto landscape as a whole: "what seems most natural to us is instead artificial, man-made, replacing what used to be ancient marshes and unrestrained streams."[3]

Taking a historical perspective, we can find similar phenomena originating from some proposed or implemented interventions to create new systems and to gradually test their effects. These are processes emerging from an impulse of innovation and that manifest themselves in a particular phase of the history of Venice characterized by processes of *renovatio* on various fronts: city, territory, and environment.[4] As Paolo Morachiello has clearly pointed out in his most recent contribution on the subject, if the dynamics of water

← *The Venetian lagoon*, second half of the fifteenth century, ASVe, *Savi ed Esecutori alle Acque*, *Disegni*, Laguna, dis. 128

movement had not been wisely governed during the modern age, if there had been no debates and consequent decisions in its defense, that is, if the forces of nature had been left free to act, the lagoon would have disappeared, becoming land or returning to the sea.[5] And the Grand Canal would be a multilane motorway with traffic lights hanging from the Rialto bridge at the crossroads with other lateral canals/roads: this was affirmed by Arnold Esch when he illustrated the intertwining of humanity, nature, and history that defines some significant historic Italian landscapes, including Venice, which "is in the water, obeys only water, dominates the water."[6]

It is in this dialectic between nature and artifice, where the overcoming of technical obstacles such as those related to building "over salt water"[7] and defending oneself from it, inspired innovative solutions over the centuries, that the history of the city of Venice is inscribed. As Élisabeth Crouzet-Pavan has shown, the centuries-long struggle against water and storm surges is one of the constituent elements of the myth of the early city. She calls it a progressive invention "à la merci des eaux," born and bred in the midst of water, mud, and reeds, in a condition of permanent difficulty.[8] Recent events have highlighted this well. A vivid testimony of how complicated it was to survey "marshes and lagoons" in order to try to find new solutions to the problems that afflicted this amphibious environment is offered to us by Fra Giocondo in the early sixteenth century. When he was asked about the diverting of the River Brenta, and more generally how "to provide for the mud and fresh waters that form land for Venice,"[9] he struggled through the reeds, observing ditches and embankments to find the old riverbed, measuring and assessing everything "with his eye, which is more certain than the ear" (which is to say, than word of mouth).[10] Finally, having collected the opinions of his predecessors, the humanist friar declared himself against forcing nature by changing the course of the rivers: instead, he recommended exploiting existing stretches of natural channels. Despite this, the tendency prevailed of relying on practical experience by means of large-scale works, in this case the diversion of rivers. A few years earlier, an anonymous surveyor had complained of "both mental and corporeal" fatigue suffered in "plucking oneself from canals, creeks, woods, scrub, swamp": a complex environment that was constantly changing.[11]

It was precisely the difficulties imposed by the environmental conditions, and the intentionality and determination to

overcome them, that created the dialectical relationship between imagination and pragmatism which, according to Deborah Howard, has characterized the entire history of the city, contributing to the formation of a "corpus of conceptual ideas that have continually stimulated debate."[12]
Ideas that stimulate debate and create a shared planning process between governing bodies, administrators, and intellectuals that is needed now more than ever. In other words, in controlling and managing the delicate and vulnerable environment, a real culture of governance is formed, as in a modern laboratory for the city and its lagoon. That history can help us understand complex systems such as those we are considering is something that has been mentioned many times in recent years: after all, Venice and its lagoon offer us the best horizon to discuss great environmental issues, which concern not only scientific knowledge, but also history and, more generally, culture.[13]

Eternal life: governing water

Water and land: the relationship between these two elements can be taken as paradigmatic of the changing city. But the entire lagoon environment, amphibious as it is, is conditioned by it. In 1715, Bernardo Trevisan chose Andrea Zucchi's engraving *Opponesi elemento ad elemento* as the frontispiece to his treatise on the lagoon. It depicts two allegorical female figures in a struggle: water and land are pitted against each other, trying to invade their respective spaces. The scene, which takes place on a beach at the edge of the lagoon, with the outline of St. Mark's Square in the background, seems to allude to the literary image that had spread two centuries earlier thanks to Marcantonio Sabellico: that of a city "suspended between two elements,"[14] in a precarious and unstable equilibrium that required the utmost attention from the governing bodies. In the inscription dictated by the humanist Giovanni Battista Cipelli, known as Egnazio, carved in Latin characters on black marble and set into the wall at the Rialto in the offices of the *magistratura* responsible for the lagoon, we again find the evocation of a Venice founded on water, surrounded and protected by the lagoon like walls.[15]
But what really makes the survival and development of a city in the middle of the water possible is the "care," the timely maintenance work carried out continuously by the *magistrature* of the Venetian State. In a context characterized in the mid-sixteenth century by one of the highest rates of urbanization in Europe, the existence/survival of the city was assured by a model of controlled and balanced management of the waters,

← Andrea Zucchi, *Opponesi elemento ad elemento*, in Bernardo Trevisan, *Della laguna di Venezia*, Venice, printed by Domenico Lovisa, 1715, frontispiece
↑ JDS Architects, Julien De Smedt, *Aqualta* 2060, digital rendering presented at the Architecture Venice Biennale 2010

perceived as both a resource and a threat.[16] In the course of the modern age, Venice and its lagoon became the object of protection, in an innovative dimension, as hydraulic issues, hygiene, and urban decorum were considered to be interdependent. Albeit clearly present at least since the end of the thirteenth century, as Crouzet-Pavan has well demonstrated, these issues accelerated considerably around the middle of the sixteenth century.[17] At the same time, there was a discussion about damming the lands that were "falling" into the water (as Daniele Barbaro, among others, has effectively shown), about contrasting the invasion of brackish waters and moving the rivers away from the lagoon closest to Venice.[18] However, it was not only a question of defending and protecting the hydraulic balance of the Venetian environment, but also of shaping the urban fabric, counteracting the progress of the land at the expense of the water environment by means of banks and embankments.[19] In other words, maintenance on an urban scale becomes a complementary field to regulating rivers and territory. The development of this approach can be followed by noting a series of decisions relating to the lagoon, the defense of the beachfronts from sea erosion, the excavation of the channels and creeks, and the maintenance of the banks, implemented by the Savi ed Esecutori alle Acque, given a permanent standing in 1505 for the hydraulic management of the urban and lagoon territory.[20] The main lines of action, which were complementary, and in relation to which the archives of the magistracy were organized, included the lagoon, beachfronts, and rivers.

While in the initial phase of activity the procedures did not appear systematic, the situation changed after 1531, when a specific resolution of the Officio alle Acque established that, in order to prevent washed-out silt from swamping the lagoon, open spaces had to be bordered with stone banks.[21]
A wide range of interventions is recorded in the archives of the *magistrature* concerning the establishment of a clear land-water border "recto tramite," the reclamation of muddy swamp, the division of new plots of land wrested from the water, and the removal of silt from the canals.[22] The inspections undertaken by the *proti* regulated and inspected the works requested by anyone wishing to transform parts of the city. One example is that of the Corpus Domini nuns, who in 1574 intended to close off their vegetable garden "which ends by the lagoon on the west side towards Santa Chiara" with a stone bank to replace the wooden fence.[23] These are documents of great

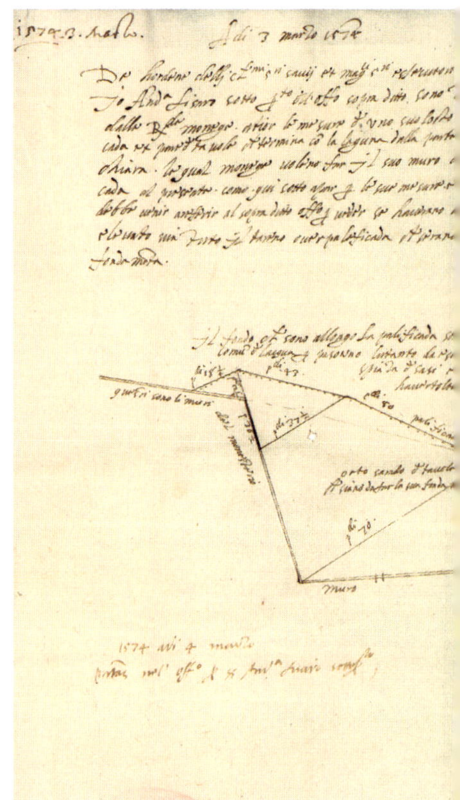

immediacy and include drawings based on a regular geometric framework, the result of a "meticulous" environmental survey.[24] The manuscript maps, such as the one of the Giudecca island drawn by Paolo da Castello in 1539 to "atterrar le sache drio la Zuecha etiam drezar la fondamenta,"[25] are very effective in illustration the process of transformation of the features characterizing the lagoon landscape.

In the deliberations, individual citizens who failed to take care of the common good were regularly called to order: "usurpations," misappropriations, and encroachments (the result of the "greedy desires of people," according to Cristoforo Sabbadino, the most famous of the *proti* working for the *magistrato*)[26] jeopardized the fragile balance of the lagoon. The intensification of operations was linked to the emergence of a new concept of urban decorum.[27] In illustrating the goals of his "plan" for Venice, the same as those of the Officio alle Acque for which he worked, Sabbadino himself presented a clear cost/benefit analysis to demonstrate how the planned actions were overall of benefit for the balance of the lagoon: "Non s'atterrerà più a torno Venetia, ma più presto si caverà, et mantenerassi canali a torno, a torno, e Venetia serà la più bella et più commoda città del mondo senza nessun danno di quella."[28] The production of new building land provides a use for excavated silt and at the same time brought considerable benefits as regarded the preservation and "cleaning" of the lagoon, but also ended up resolving problems of urban growth and environmental protection in an overall perspective. The vision of the engineer and proto of the Officio alle Acque is an organic vision typical of the Venetian *magistratura*, which implemented a very modern environmental policy suggested by the limits of the natural resources and by wisdom in their administration,[29] based on the correct use of water and land, but also on the use/consumption of land, principles similar to those found in today's striving for "calculated harmony" with regard to the built landscape.[30]

Creating land and construction of the lagoon city

In the "most noble and singular city," which in the mid-sixteenth century had to respond to major demographic growth, producing new building land meant reclaiming, cleaning, and recovering marshy lands from the lagoon, "making up for nature's defects with artifice," states Francesco Sansovino.[31] With the main part of the city now congested and subject to a very high population density, the focus shifted to the outlying areas. But any action arising in the context of a fragile equilibrium needs to anticipate its effects. It has been

↑ Andrea Fisaro, ground plan of the vegetable garden belonging to the Corpus Domini nuns and surrounding area, March 3, 1574, ASVe, *Savi ed Esecutori alle Acque*, b. 530, dis. 3
→ Map of Cannaregio between San Felice, the Servi, Madonna dell'Orto, Misericordia, and Santa Caterina, 1530s, ASVe, *Savi ed Esecutori alle Acque*, Disegni, Laguna, Diversi 128/7

LI SERVI
PROCVRATIA
CONTARINI

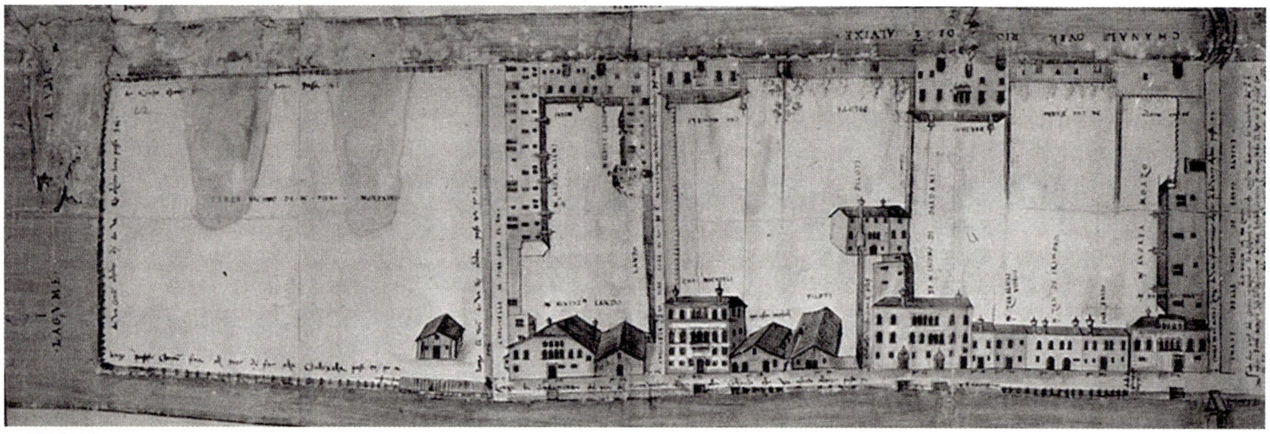

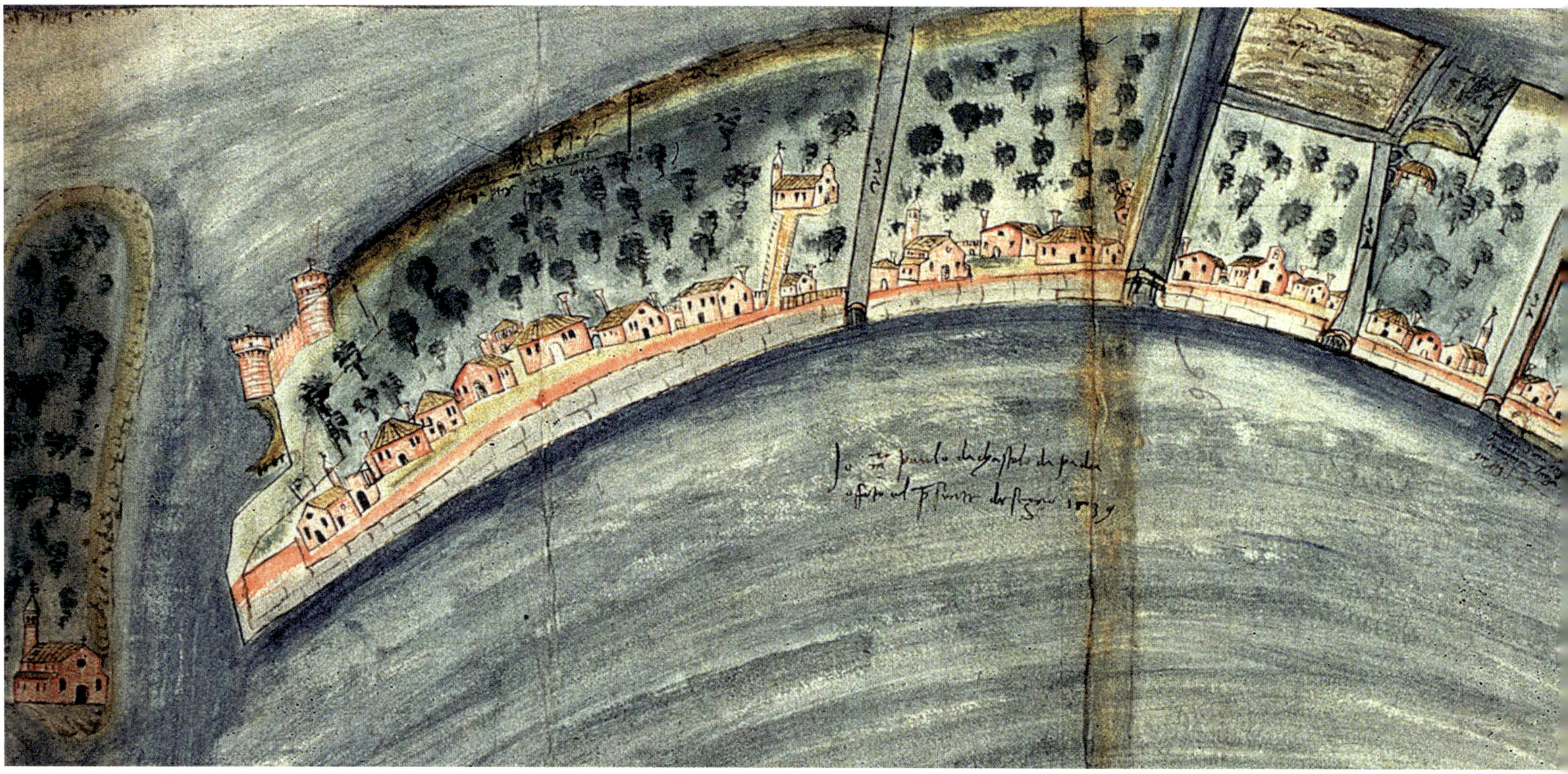

↑ Nicolò dal Cortivo, perspective drawing of the houses belonging to the nuns of Sant'Alvise on the canal of the same name with the delimitation of the "teren vachuo" (empty plot) by Piero Morosini, January 30, 1538, ASVe, *Miscellanea Mappe*, 1216
↑ Paolo da Castello, *Pianta dell'isola della Giudecca*, 1539, ASVe, Savi ed Esecutori alle Acque, Disegni, Laguna

pointed out how the "great and precocious environmental sensitivity" presents "clear points of contact" with the medical thinking of the time.[32] There is no substantial agreement on this point, however. For his part, Andrea Zannini fears the existence of an "eco-myth" regarding the environmental orientation of the aristocratic government: an "assumed" environmental sensitivity, in that it applied above all, or exclusively, when the interests of the Venice lagoon were at stake.[33]

Be that as it may, the historical documentation shows how the organs of state took on costs and decisions, matured through listening to and comparing experts in the field. At the center of the activity were suitably trained professionals, the proti, capable of moving between different disciplines—hydraulics, geography, cartography, and architecture. An examination of the organizational chart of the Magistrato alle Acque shows that it included a sort of hydraulic school where "aspiring *proti*" could complete their apprenticeship by working alongside more experienced and mature engineers. In order to support the work of environmental and urban maintenance, it was necessary to have surveys and measurements, but also to carry out inspections to verify their accuracy and to intervene in the case of misappropriation of land. In the archives of the Savi ed Esecutori alle Acque there is a large collection of drawings of the "making of the city":[34] characterized by practical purposes, as a means of controlling physical transformations, they constitute the most significant product of the activity of the *magistrati* and *proti* in their service, particularly in the outlying areas, where the land–water relationship required constant attention.

In 1642, Bernardino Contin drew up the "indice di dissegni di laguna, Pò, Brenta, e Piave, dissegni diversi, e spolveri essistenti nel magistrato eccellentissimo alle acque."[35]

The first complete inventory of which we are aware comprised no fewer than five hundred illustrative documents, some of which have now been lost or incorporated into the archives of other *magistrature*.[36] Among the projects elaborated it is possible to identify some on a large scale, ambitious in their objectives; others outlining small areas of town threatened by encroaching water; others still relating to the official activity aimed at verifying the progression of the city's margins at the expense of the water.

The two scales of response—planned and emergency measures—are not contradictory but belong to a system engrained in the Venetian administration. Wide-ranging

↑ Cristoforo Sabbadino, map of the lagoon from San Pietro in Volta to Sant'Erasmo, ASVe, *Savi ed Esecutori alle Acque*, Disegni, Laguna, diversi 128/3
→ Cristoforo Sabbadino, *Pianta de Venetia*, 1557, ASVe, *Savi ed Esecutori alle Acque*, Disegni, Laguna

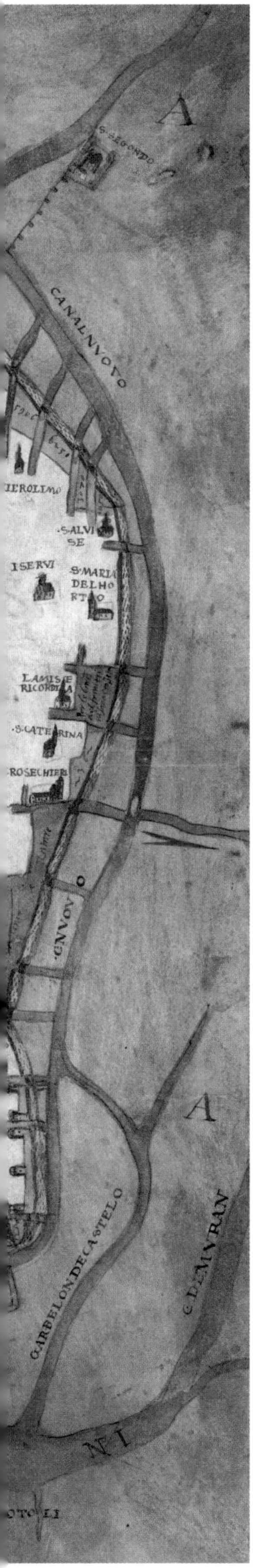

programs are intertwined with a day-to-day policy, made up of small-scale responses that, as a whole, make up a continuous, even ordinary, maintenance. Drawings, maps, and “amphibious” charts well represent that kind of modern-era documentation interpreted by Emanuela Casti as “a tool for implementing and proposing territorial strategies.”[37] It is not surprising in this context, which involves not only the management of works but also the prediction of outcomes, to find the most ambitious urban “plan” ever produced in the Italian peninsula during the Renaissance. This is the Pianta de Venetia produced by Cristoforo Sabbadino in his capacity as “*ingegnero et proto dell'uffitio dell'acque*.”[38] Dated 1557, at a time of great interest for environmental policies, it is the result of a wide-ranging reasoning on the profile and nature of the city, and ambitious in its attempt to tackle various problems (hydrographic equilibrium, regulation of the canals, protection of the lagoon and the city) by proposing solutions on a single sheet, on both an urban and territorial scale. Sabbadino's *Venetia* is in fact framed by the lagoon with its main canals, of ancient fluvial origin or induced by tidal penetration. The focus is on the outlying areas, on the “empty” lands, the grassy fields, and the muddy, low-lying areas that we see depicted fifty years earlier in Jacopo de' Barbari's bird's-eye view of Venice. Once reclaimed, the marginal lands would be given to the city in the form of spaces to be parceled out and put on the building market or, alternatively, used for other facilities (boatyards, docks, bridges). According to his calculations, reported in the margins of the drawing, the sale of the new lots wrested from the water would have made it possible to cover the costs of both the construction of the long stone bank and the bridges that would have assured access throughout the length, along with the excavation of perimeter canals, by adopting a system we would now call project financing. This would have had a positive effect on the overall hydrodynamics of the lagoon: the production of new building land, obtained through filling by reusing the silt removed from the canals, would ultimately have led to better water circulation. Finally, it would have had a positive effect on the “beauty” of the city.
Although there are no certain data on the commissioning of the drawing, it adheres to the practical aims of the office, to the requests of the *magistrati* to whom it was evidently addressed, and of whom Sabbadino was the operational arm. In 1557, the year the plan was drawn up and three years before his death, the proto Sabbadino, a native of Chioggia, wrote

a sort of testament, setting out the main issues he had tried to resolve in a number of "memoirs." As is well known, the aim was an overall project to restore the original conditions of the lagoon, corrupted by external agents (the rivers, the sea, and men, according to his well-known sonnet).[39] His approach, as opposed to that of Alvise Cornaro, promoter of the reclamation of vast areas from the lagoon, has been defined as a conservative approach in order to ensure that as much as possible of the water basin would be left untouched. Underlying this was the idea that the "conservation of the lagoon" represented the preservation of the "fortress of Venetia." "A voler che questo corpo si conservi vivo, bello, sano e gagliardo, fa bisogno conservarlo tutto intiero,"[40] he stated in his writings, proposing an organic vision of the lagoon, comparing it to a human body that breathes through the ebb and flow of the lagoon water. Effectively illustrated in drawings through the skillful use of colors and evoking a lagoon landscape suspended between land and water, the anthropomorphic metaphor seems to echo Fernand Braudel's words. When we look at Venice, made to be understood and loved by all, says the great French historian, it is like observing "a swimmer lifting his head out of the water."[41]
Repeatedly put to the test, not least because of climate change, the survival of Venice has returned to center stage in the pandemic, which has offered us persuasive testimony of its fragile beauty, but which has also brought to light the difficulties and obstacles of everyday life. Its salvation has been discussed at least since the *aqua granda* of 1966, sometimes invoking protection in the face of the "miracle" that still keeps it alive, in tones that are really not so different to Sabbadino's *Discorsi sopra la laguna*.[42] Making the city an exemplary laboratory, precisely on the basis of the centuries-old "struggle" to ensure its survival, is an appeal that recurs in the pages of those who have dealt with Venice and its waters from various points of view.[43] All that remains is to reiterate it with an eye to the numerous problems of the city and its lagoon today, together with the most extraordinary of "ecological fairy tales" that can be told to new generations.[44]

1. Francesco Erbani, *Non è triste Venezia. Pietre, acque, persone. Reportage narrativo da una città che deve ricominciare* (San Cesario di Lecce: Manni, 2018), 46–47.
2. Élisabeth Crouzet-Pavan, *Venise: une invention de la ville. XIIIe-XVe siècle* (Syessel: Champ Vallon, 1997); Joanne M. Ferraro, *Venice. History of the Floating City* (Cambridge: Cambridge University Press, 2012).
3. Guido Piovene, *I saggi*, edited by Luciano Simonelli, vol. II (Milan: Mondadori, 1990).
4. Ennio Concina, "Ampliar la città: spazio urbano, 'res publica' e architettura," in *Storia di Venezia*, vol. VI, *Dal Rinascimento al Barocco*, edited by Gaetano Cozzi and Paolo Prodi (Rome: Treccani, 1994), 253–273.
5. Paolo Morachiello, "Vita e sopravvivenza della laguna nel 'dialogo' tra due cittadini della Repubblica nel XVI secolo (e il suo seguito negli anni successivi)," in *Acqua e cibo a Venezia. Storie della laguna e della città*, exhibition catalogue (Venice, 26 September 2015–14 February 2016), edited by Donatella Calabi and Ludovica Galeazzo (Venice: Marsilio, 2015), 103–107. See Luigi D'Alpaos, and A. Rinaldo, "Forma e funzione della laguna," *Acqua e cibo*, 35–38; Andrea Rinaldo, *Il governo dell'acqua. Ambiente naturale e ambiente costruito* (Venice: Marsilio, 2009).
6. Arnold Esch, *Viaggio nei paesaggi storici italiani* (Gorizia: LEG, 2020), 65.
7. Élisabeth Crouzet-Pavan, "Sopra le acque salse": *espaces, pouvoir et société à Venise à la fin du Moyen Âge* (Rome: École française de Rome, 1992), 2 vols.
8. Crouzet-Pavan, *Venise: une invention de la ville*, 10; Ermanno Orlando, "Venezia e la laguna," in *Paesaggi delle Venezie. Storia ed economia*, edited by Gian Pietro Brogiolo, Andrea Leonardi, and Carlo Tosco (Vicenza–Venice: Istituto per le ricerche di storia sociale e religiosa–Marsilio, 2016), 343–348.
9. Comment by Marin Sanudo, quoted by Giangiorgio Zorzi, "Notizie di arte e di artisti nei Diarii di Marin Sanudo," in *Atti dell'Istituto Veneto di Scienze, Lettere ed Arti*, Classe di scienze morali e lettere, vol. CXIX, 1960–1961, 471–604, esp. 554.
10. A copy of the "deposition de fra Zuane Iocondo" (1506) is preserved in Venice, Archivio di Stato (hereinafter ASVe), *Savi ed Esecutori alle Acque*, reg. 161, c. nos., and has been transcribed in Bernardino Zendrini, *Memorie storiche dello stato antico e moderno delle lagune di Venezia*, vol. II (Padua: Stamperia del Seminario, MDCCCXI), 247–262; Vincenzo Fontana, "Fra Giovanni Giocondo ingegnere idraulico a Venezia (1506–1508)," in *Giovanni Giocondo umanista, architetto e antiquario*, edited by Pierre Gros and Pier Nicola Pagliara (Venice: Marsilio, 2014), 363–379.
11. Eugenia Bevilacqua, "Notizie preliminari intorno a documenti di rilevamento cartografico del territorio veneziano della fine del XV secolo," in *Atti dell'Istituto Veneto di Scienze, Lettere ed Arti*, Classe di scienze morali e lettere, vol. CXXIX, 1970–1971, 143–151, esp. 147.
12. Deborah Howard, *The Architectural History of Venice* (New Haven & London: Yale University Press, 2002), 295; Ead., *Venice disputed: Marc'Antonio Barbaro and Venetian Architecture, 1550–1600* (New Haven & London: Yale University Press, 2011).
13. Amitav Ghosh, *La grande cecità. Il cambiamento climatico e l'impensabile* (Vicenza: Neri Pozza, 2017).
14. Elena Svalduz, entry in *Acqua e cibo a Venezia*, 250.
15. Salvatore Settis, Premessa, in *Acqua e cibo a Venezia*, 19–23.
16. Franco Cazzola, "L'incubo delle acque nel Cinquecento," in *Verso la santa agricoltura. Alvise Cornaro, Ruzante, il Polesine*, edited by Gino Benzoni (Rovigo: Minelliana, 2002), 59–65.
17. Among the other contributions, Élisabeth Crouzet-Pavan, "Venice and Its Surroundings," in *A Companion to Venetian History, 1400–1797*, edited by Eric R. Dursteler (Leiden & Boston: Brill, 2013), 25–46.
18. Giuliana Mazzi, "Cartografia e organizzazione del sapere tecnico," in *Il governo delle acque*, edited by Maria Francesco Tiepolo and Franco Rossi (Venice: Istituto Veneto di Scienze, Lettere ed Arti, 2008), 51–68, esp. 51.
19. Ludovica Galeazzo, *Venezia e i margini urbani. L'insula dei Gesuiti in età moderna* (Venice: Istituto Veneto di Scienze, Lettere ed Arti, 2018).
20. Elena Svalduz, "Al servizio del magistrato. I proti alle acque nel corso del primo secolo d'attività," in *"Architetto sia l'ingegniero che discorre." Ingegneri, architetti e proti nell'età della Repubblica*, edited by Giuliana Mazzi and Stefano Zaggia (Venice: Marsilio, 2004), 233–268; Silvia Gasparini, "Le procedure amministrative e giudiziarie," in *Il governo delle acque*, 87–152; Giovanni Caniato, "Tra dolce e salto," in *Acqua e cibo a Venezia*, 31–38.
21. ASVe, *Savi ed Esecutori alle Acque*, reg. 333, c. 9v, November 6, 1531.
22. Elena Svalduz, "Visti dall'acqua: i disegni del 'far la città' e la manutenzione urbana," in *Fare la città. Salvaguardia e manutenzione urbana a Venezia in età moderna*, edited by Stefano Zaggia (Milan: Bruno Mondadori, 2006), 71–96.
23. ASVe, *Savi ed Esecutori alle Acque*, b. 530, dis. 3, March 3, 1574, drawing by Andrea Fisaro, *sottoproto* of the *Officio alle Acque*.
24. Barbara Mazza, "Politica lagunare di Venezia nel Cinquecento e interventi sul territorio: note di cartografia," in *Architettura e utopia nella Venezia del Cinquecento*, exhibition catalogue (Venice, July–October 1980) (Milan: Electa, 1980), 130–143.
25. "Fill in the basins behind the Giudecca and also straighten the banks." Svalduz, *Al servizio del magistrato*, 260–261; see Filippo Toso, "Paolo da Castello 'proto et inzegnero,'" in *"Architetto sia l'ingegniero che discorre"*, 269–306, esp. 298–299.
26. Cristiano Tentori, *Della legislazione veneziana sulla preservazione della laguna. Dissertazione storico-filosofico-critica* (Venice: Giuseppe Rosa, 1792), 137.
27. Manfredo Tafuri, *Venezia e il Rinascimento. Religione, scienza, architettura* (Turin: Einaudi, 1985), 156–169.
28. "The lagoon will no longer silt up around Venice, but will soon be excavated, and with the canals being each in turn maintained, Venice will be the most beautiful and commodious city in the world without any loss or impairment to it." ASVe, *Savi ed Esecutori alle Acque*, Disegni, Laguna, dis. 14. Regarding Sabbadino's city plan, Elena Svalduz, *Venice 1557: Sabbadino's City Plan*, in *Architecture, Art and Identity in Venice and its Territories, 1450–1750. Essays in Honour of Deborah Howard*, edited by Nebahat Avcioğlu and Emma Jones (Farnham: Ashgate, 2013), 71–86.
29. Luigi D'Alpaos, *Fatti e misfatti di idraulica lagunare. La laguna di Venezia dalla diversione dei fiumi alle nuove opere alle bocche di porto* (Venice: Istituto Veneto di Scienze, Lettere ed Arti, 2010).
30. *Il paesaggio costruito, il paesaggio nell'arte*, edited by Gianmario Guidarelli and Elena Svalduz (Padua: Padua University Press, 2017).
31. Francesco Sansovino, *Venetia città nobilissima et singolare*, with extra material by Giustiniano Martinioni, Appresso Steffano Curti, Venetia MDCLXIII, 381.
32. Luca Ciancio, s.v. "Sabbadino, Cristoforo," in *Dizionario Biografico degli Italiani*, vol. 89, 2017, www.treccani.it/enciclopedia/cristoforo-sabbadino_%28Dizionario-Biografico%29/.
33. Andrea Zannini, "Un ecomito? Venezia (XV–XVIII sec.)," in *Storia economia e ambiente italiano (ca. 1400–1850)*, edited by Guido Alfani, Matteo Di Tullio, and Luca Mocarelli (Milan: Franco Angeli, 2012), 100–113.
34. Svalduz, *Visti dall'acqua*.
35. Venice, Biblioteca del Museo Correr (hereinafter BMCVe), Ms Pd 8b.
36. For example, Nicolò dal Cortivo (ASVe, *Miscellanea Mappe*, 1216) is marked "20" on the rear, a number corresponding to the inventory number: "20. Dessegno del Rio de San Alvise del sopraddetto 1538 30 zener."
37. Emanuela Casti, *L'ordine del mondo e la sua rappresentazione. Semiosi cartografica e autoreferenza* (Milan: Unicopli, 1998), 77–98.
38. Svalduz, *Venice 1557*, with earlier bibliography.
39. Tafuri, *Venezia e il Rinascimento*, 215–218; Salvatore Ciriacono, *Acque e agricoltura. Venezia, l'Olanda e la bonifica europea in età moderna* (Milan: Franco Angeli, 1994), 138–158.
40. "In order for this body to be kept alive, beautiful, healthy, and vigorous, it is necessary to keep it whole." Tafuri, *Venezia e il Rinascimento*, 237; Cristoforo Sabbadino, *Il sistema Laguna a metà Cinquecento. Opere scelte pubblicate nel 450° della morte*, edited by Pier Giorgio Tiozzo Gobetto, with introduction by Salvatore Ciriacono (Sottomarina: Il Leggio, 2011).
41. Fernand Braudel, *Venezia* (Bologna: il Mulino, 2013), 23.
42. Cristoforo Sabbadino, *Discorsi sopra la laguna*, with editing by Bernardino Annibali (1602), BMCVe, Ms. Donà dalle Rose, n. 86, c. 86v: the "Divine City" was born from divine majesty, who "placed and built it in the water, built the walls of water, made the gates and streets of water, in order that to maintain the walls, towers, streets, and gates they do not need stone, lime, wood, or iron, but only faith, prudence, and justice, of which this divine city has never lacked (by the grace of God)."
43. For example, Eugenio Vassallo, "Telematica e beni culturali: un'ipotesi per il futuro che parte da Venezia," in *A vent'anni dall'evento di marea del novembre 1966*, proceedings of the study day (November 3, 1986) (Venice: Istituto Veneto di Scienze, Lettere ed Arti, 1987), 77–81.
44. Piero Bevilacqua, *Venezia e le acque. Una metafora planetaria* (Rome: Donzelli, 1998), 13.

The MOSE system

Consorzio Venezia Nuova

MOSE

← The MOSE Experimental Electromechanical Module: the full-scale prototype of a floodgate used to optimize the operation of the system, testing its components, construction materials, and their reaction to the aggression of natural elements. It consisted of a floodgate mounted on a hull, surmounted by a structure housing the control center and technical rooms.
↓ The module built in Porto Marghera was transported to the Lido Treporti canal where it was tested from 1988 to 1992.

All the activities falling within the State's jurisdiction concerning the physical and environmental protection of Venice and its lagoon are based on a General Plan of Interventions (Piano generale degli interventi) linking the protection of the territory from flooding and the safeguarding of the ecosystem, and making use of distinct but mutually related lines of action: defense from *acque alte* (high waters), environmental defense, and defense from sea storms. This General Plan of Interventions, which is referred to in the special legislation for Venice, is defined by Italian law 798/1984 as the Joint Committee for Policy, Coordination and Control (Comitato misto di indirizzo, coordinamento e controllo—the so-called "*Comitatone*," or "megacommittee"), where the competent national and local institutions are represented. The MOSE system as a whole is the integrated plan of measures affecting the entire lagoon area (550 km^2 in size) and has been implemented to respond to all the critical elements of crisis.
Starting with the need to create a secure, continuous line of defense against the threat from the sea, achieved by building barriers at the lagoon entrances and completely reinforcing the coastline, the project simultaneously worked to encourage the development of natural dynamics and habitats within the lagoon basin responsible for the hydrodynamic and ecological functions essential for a healthy ecosystem. At the same time, action has been taken on the edge of the lagoon (particularly in Porto Marghera) to block sources of water and seabed pollution. The occurrence of the high water phenomenon is mainly linked to two causes that influence sea level on different time scales: the temporary rise in sea level due to the combined effect of the astronomical tide and the meteorological component and the relative lowering of land compared to the average sea level, due to two factors acting on a long time scale: subsidence (lowering of land level) and eustasy (rising of the sea level).
Subsidence in the lagoon area has both a natural and a manmade cause, consisting mainly of the exploitation of groundwater resources for industrial purposes since the 1930s, which considerably worsens the former cause. The extraction of groundwater also

results in a decrease in pressure in the subsoil and therefore its compaction. The most critical phase of this process began in the 1950s and worsened until 1970. Eustasy is linked to global climatic changes. During cold periods, precipitation is retained on the continents in the form of ice and consequently the sea level drops; the opposite occurs during warm periods. In the future, the high water phenomenon will be exacerbated by the expected rise sea levels as a result of climate change. With respect to this problem, MOSE (together with the reinforcement of the coastal strip) has been designed in line with a precautionary criterion to cope with a sea level rise of up to 60 cm, which is higher than even the most recent estimates of the Intergovernmental Panel on Climate Change (IPCC), which predict a sea level rise in the next 100 years in a range between 18 and 59 cm. Large areas of the Italian coastline will be at risk of flooding as a result of the rise in the Mediterranean Sea due to both climate change and geological features: according to an ENEA estimate, tens of square kilometers of land will be lost by the end of the century.

Physical models, prototypes, and tests

In the aftermath of the *aqua granda*, or exceptionally high tidal flood, of November 4, 1966, the numerous crises afflicting the lagoon area became evident. The emergency of that dramatic day was the most serious symptom of a dramatically increasing trend in the frequency of floods. Over the course of the twentieth century, the city's age-old relationship with water changed as a result of rising sea levels and the lowering of the land, mainly because of human activity, and these factors

SQUALUS

caused the phenomenon of high water to increase in frequency and intensity. Moreover, the ecosystem was already showing other signs of weakness at that time, ranging from the fragility of the coastal territory, increasingly exposed to storm surges, to the phenomenon of erosion, with the progressive disappearance of morphological structures and the loss of their fundamental hydrodynamic functions. In addition to all this, there was also the problem of pollutant spillage in some areas.
After the 1966 flood, the need to protect the city from high water began to be discussed. Law no. 171/1973 for the protection of Venice marked the beginning of a long process of studying the possibilities to protect the city and the lagoon from high water, and in 1975 the Ministero dei Lavori Pubblici (Ministry of public works) launched a competition to find solutions to this problem.
Five groups of companies took part in the competition. In general, all the projects proposed were based on the indications contained in the "Government guidelines for drafting the district plan for Venice and its hinterland," issued by the Council of Ministers on March 27, 1975. The "Guidelines" not only established the adoption of the competitive tender procedure, but also specified that "the preservation of the hydrogeological balance of the lagoon and the reduction of high waters in the historic centers within limits such as not to disturb the functionality of the port system and the activities of the population must be achieved by means of a system of fixed works to regulate the entrances which may subsequently be supplemented by maneuverable parts, should it become

↑ Experimental Center for Hydraulic Models of Voltabarozzo, Padua. At the Centro del Magistrato alle Acque (Venice water authority centre, now the Provveditorato Interregionale per le Opere Pubbliche del Triveneto, or Triveneto interregional superintendency for public works), the behavior of floodgates was studied with physical models at different scales. In addition, a general physical model of the lagoon, physical models of the three lagoon entrances, a basin, and two canals were used for functional tests on the design of mobile works, littoral reinforcement works, restructuring of breakwaters, and reconstruction of salt marshes.

necessary—in relation to tidal levels—to close the entrances completely."
The design drafts for each applicant therefore provided for a series of fixed works at the entrances (with different configurations, extensions, or structural characteristics), supplemented by mobile devices for the progressive closure of the passages between the sea and the lagoon. The devices might be flap gates or large metal panels fixed to the seabed or "soft" systems of rubber or plastic material.
None of the five projects presented proved to be adequate. Therefore, with Decree Law 4/1980, later converted into law, the government ordered the acquisition of all the proposals so that they could be studied in order to formulate a new project that would take into account the various aspects of the "lagoon problem." Specifically, it would be necessary to carry out a feasibility study and a general project of the works necessary "to preserve the hydrogeological balance of the lagoon and to reduce high waters in the historic centers."
The result of this study was a project, called "*Progettone*" (Megaproject), which envisaged various types of work on the banks, the sea bed and the lagoon entrances. The "*Progettone*" was officially presented in 1981 and approved by the Consiglio Superiore dei Lavori Pubblici (CSLP, or High council for public works) in 1982, as was the idea of creating a defense against high waters by means of mobile works at the lagoon entrances, albeit with limitations.
The work carried out in the following years, during the development of the preliminary outline phase, included an in-depth verification of the possibility of defending Venice from flooding using various types of devices, including mobile works at the entrances and interventions to eliminate the effects produced in the lagoon by dredging the canals or building local defenses in the built-up areas with "insulae" type interventions.
Law no. 798/1984, unanimously approved by Parliament, established the absolute necessity of "protecting urban lagoon settlements from exceptional high waters" and indicated the criteria to be adopted in selecting works and their design, which remained unchanged throughout the development of the project. The law specifies that water defense must be implemented by means of mobile structures at the lagoon entrances, adopting the solution studied for the Ministero dei Lavori Pubblici in 1981. The criteria for studying these solutions and comparing them are those referred to in the 1984 law, which states that any interventions must not divide the lagoon or produce negative effects on its environment, which is already in a critical condition; this includes avoiding increasing erosion and flattening the lagoon bed or worsening the water or sediment quality. Based on the extensive knowledge of the lagoon environment, this criterion rules out any permanent modification of the tidal flows in the lagoon. In the case of mobile works at the lagoon entrances, intermediate piers, overhead superstructures, and structures in general whose characteristics and size would have a serious environmental impact are not permitted. The same applies to the need for these works not to constitute a hazard or damage to fishing and to ensure the continuity of transit through the entrances for emergency vessels during closure of the gates.
Lastly, the Superintendency stipulates that any local defenses in the historic center of Venice must take into account

the historic and monumental value of the city and of individual buildings, setting the maximum size of rises to 10–20 cm, depending on the context.
In May 1985, the "*Comitatone*" (in which the competent national and local institutions were represented) entrusted the Consorzio Venezia Nuova, which at that time consisted of twenty-seven companies, with all the work necessary to safeguard Venice in a "single concession."
The years 1989–1992 were devoted to the development, at the level of general outline, of the solution indicated in the preliminary phase and to the sizing of the structures and installations. In order to provide backing for the design and to meet the requests made during the approval process of the project drawings, more than one hundred studies were carried out on numerous subjects: hydrodynamics, the morphology of the lagoon, the coastline and entrances, water and sediment quality, weather and meteorological conditions, geotechnics, the behavior of the mobile structures, the port system, and fishing.
The effects were studied of what might be obtained with interventions to restore the morphological situation of the lagoon existing at the beginning of the last century: that is, before the reclamation of large areas and excavation of the broad lagoon canals for the development of the industrial area and port.
The possibility of achieving a defensive system through local protection by raising the margins of the *calli* and *campi* of the built-up areas was also studied. Finally, the effects produced by the movable structures of the 1981 "*Progettone*" solution, characterized by a permanent reduction in tidal flows, were studied, as well as the effects that could be obtained again with movable structures, but also evaluating a solution that would not modify the normal tidal system in any way except for short periods when the gates were closed. It was also hypothesized that each of these interventions could be carried out individually or jointly with others to achieve the best possible result. Sophisticated mathematical models were used to reproduce the hydrodynamics of the lagoon and to determine the effect produced by the various measures; each approach was always studied using more than one model to ensure that the results obtained were reliable. In the study of the movable works, each check was carried out using two- and three-dimensional physical models, hydrodynamic, morphological, ecological, and ship traffic simulation mathematical models, as well as a full-scale prototype of the type of barrier to be used. Numerous types of floodgates were analyzed and studies have shown that water buoyancy gates are the only type that meet the requirements of Law 798/1984.
The behavior of the floodgates was studied using traditional calculation methods, mathematical models, and, above all, by carrying out numerous tests with small-scale physical models and with a specially constructed full-scale model called the "Modulo Sperimentale Elettromeccanico" (Experimental electromechanical module, or MOSE).
Tests were also carried out on the wear of the moving parts along with analyses of the extent of aggression suffered by the materials from the marine environment, including the flora and fauna living in it.
Among the tests using small-scale physical models, a first series—carried out at Estramed, a center for hydraulic

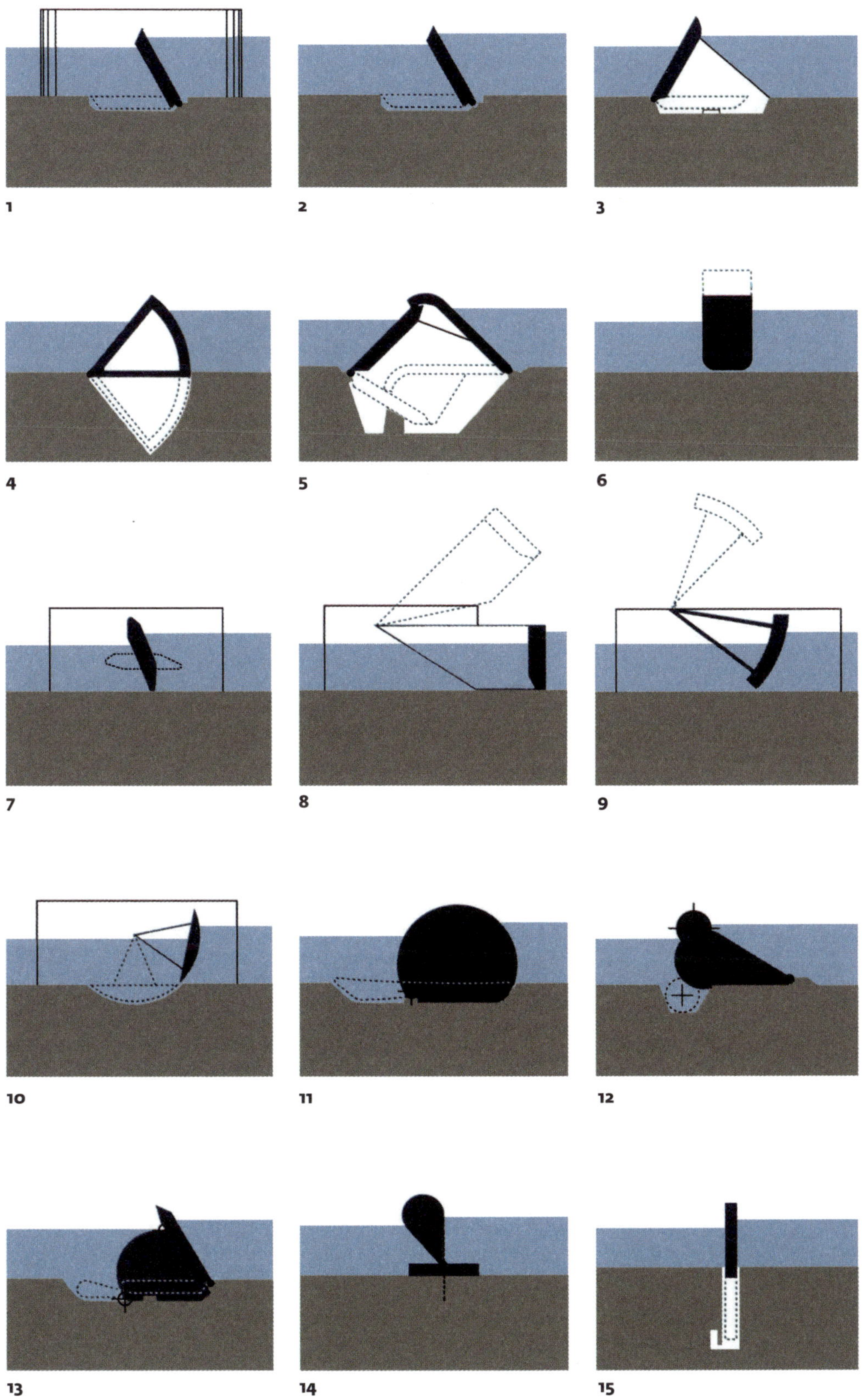

The types of gates analyzed

1. fixed straight buoyant flap gate
2. oscillating straight buoyant flap gate
3. reversed buoyant flap gate
4. drum gate
5. fish-belly flap gate
6. sliding gate
7. butterfly gate
8. dropping gate
9. sector gate
10. retractable sector gate
11. air cushion gate
12. floating gate
13. blowing gate
14. horizontal movement gate
15. emerging caisson gate

↑ → The MOSE Experimental Electromechanical Module. Different phases of the testing with trials of the movement of the floodgate.

and maritime models based in Pomezia (Rome)—was essentially dedicated to verifying the feasibility of the system. Both in-tank tests of an array of 18 floodgates and in-channel tests of 3 side-by-side floodgates hit by regular and irregular waves were carried out.
The results made it possible to introduce a first level of optimization of the dimensions of the gates into the design and to proceed with the realization of a second series of instrumental tests carried out in the Netherlands at the Delft Hydraulics Institute. Like the previous tests (with tanks and channels) but more sophisticated thanks to the larger dimensions, these made it possible to reproduce in detail the conditions in which the gates would have to operate in a real-world situation.
On the basis of the results, the preliminary design of the mobile works to regulate tidal flows in the lagoon was drawn up and the physical models required for the tests to determine the final dimensions of the gates were prepared. In this case, the models reproduced the lagoon entrances and the mobile works in the location envisaged in the preliminary design. A 1:64 scale model of the Lido entrance and two 1:60 and 1:80 scale models of the Malamocco entrance were also developed.
At the end of the deliberations of the Commissione di Esperti Internazionali (Commission of international experts), the experimentation was then integrated by examining the floodgate envisaged in the design for the Chioggia entrance. Due to its natural geographical and bathymetric conformation, this is the lagoon entrance most exposed to wave action. Two simplified models were therefore produced on a scale of 1:10 and 1:30, reducing the row to 5 gates instead of 18—as envisaged in the project—for reasons of size, flanked at the ends, for symmetry, by two gates of half the width.
As already mentioned, a full-scale model—MOSE—which was operational in the Treporti canal from 1988 to 1992, was also built. This was a full-scale reproduction of a flap gate, the hinge-connector systems, the shock absorbers, the systems for removing water from the flap gates by means of compressed air, the systems for adjusting and measuring their angle of tilt during operation, the protective paints, and the systems for removing any sand that might be deposited in the compartment housing the flap gate when at rest.
The module was inaugurated and lowered into the water on November 3, 1988. It consisted of a huge underwater metal caisson (20 × 17.5 m) mounted on a hull (32 × 25 m) topped by four 20-m-high columns and a crane. As this was a single floodgate and not an array built up to the two banks of the canal, as would be the case with the completed defense works, this module was also equipped with service facilities such as command and control rooms, offices and staff accommodation. The MOSE was towed to the Lido Treporti breakwater by two tugs.
In the first phase of MOSE, completed in 1989, a solution for the operation of the floodgate was tested, which included equipment being placed within the gate itself. The reliability analysis of the system showed that functionality could indeed be achieved with this solution, but would require as many as seventy-nine interventions per year to replace faulty components. The special components remaining on the gates in the general design are the hinge-connector assemblies, the instruments

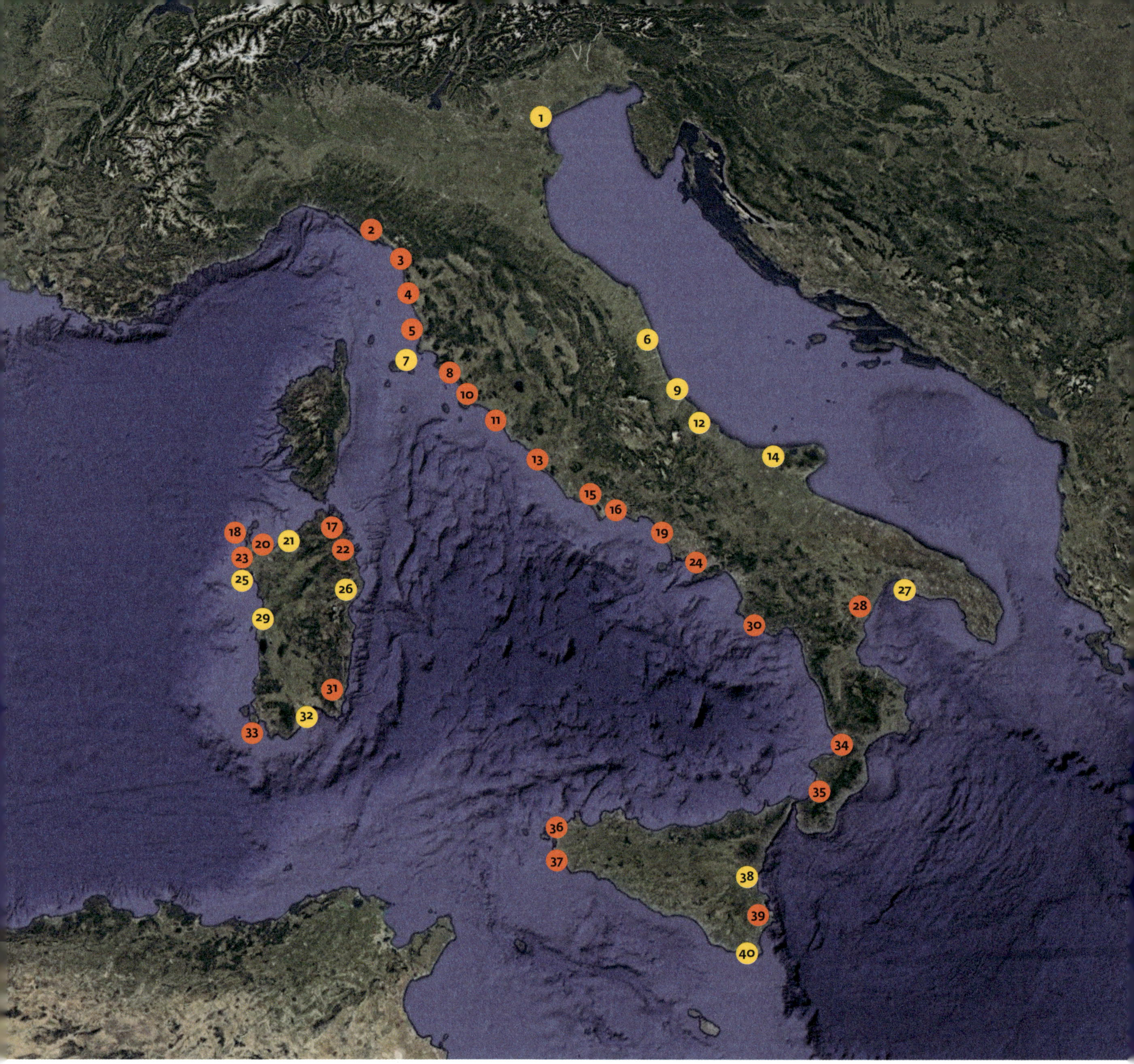

40 coastal areas at risk

completed maps

maps to be completed

1. Po Valley
2. La Spezia
3. Versilia and Massa
4. Cecina
5. Follonica and Piombino
6. Tronto river mouth
7. Marina di Campo, Elba
8. Grosseto
9. Sangro river mouth
10. Albinia
11. Burano Lake
12. Pescara river mouth
13. Plain of river Tevere
14. Lesina Area
15. Pontina plain
16. Fondi plain
17. Porto Pollo
18. Nodigheddu
19. Plain of river Volturno
20. Platamona
21. Valledoria
22. Olbia
23. Pilo
24. Plain of river Sele
25. Fertilia
26. Orosei
27. Taranto
28. Metaponto
29. Oristano
30. Alento
31. Flumendosa
32. Cagliari
33. Porto Pino
34. Santa Eufemia
35. Gioia Tauro
36. Trapani
37. Marsala
38. Pantano Longarini
39. Noto
40. Granelli

for detecting and controlling the angle of the gates, and the dampers.
The first assembly tested on MOSE was produced in the US by a company with twenty years' experience in this specific field. On the basis of the results of this experimentation, two other hinge-connector assemblies have been studied, both of Italian manufacture, with more specific characteristics for application to the gates, improved with respect to the previous assemblies in terms of the tolerances allowed for in effecting the connection with the housing structure, the time required to perform this operation and the reliability of the joint. The other components that remain attached to the gate are the dampers, which absorb the residual energy of the gate when it returns to its rest position in the housing (known as the recess) during the lowering maneuver. Various alternative solutions were also tested on MOSE for these components: hydraulic dampers with inert gas operation and more current dampers made of polymers with varying deformation under load. On the basis of the results of the experiments, the solution consisting of two pairs of rubber shock absorbers of the type normally used as shock absorbers for ships during quayside approach maneuvers was adopted in the preliminary design. At the design level, the development of the system of flap gates for the barriers and of the other interventions to be carried out and integrated also took into account the aspects of subsidence and eustasy, as explained below.
The design of the barriers, 78 floodgates and associated works of the MOSE system was conceived and developed by Technital spa, under the guidance and responsibility of its chairman Alberto Scotti. This is one of Italy's oldest and largest engineering companies, founded in 1964 and specializing in engineering services for highly complex works, and occupies a prominent international position. The design of these works was carried out with the collaboration of numerous specialized institutes and national and international experts who, under the coordination of the designer, carried out countless tests on physical models, studies with mathematical models, and measurement campaigns.

The configuration of the barriers

For each lagoon entrance (Lido Treporti, Lido San Nicolò, Malamocco, and Chioggia), the planned works consist of the installation of mobile devices (called flap gates) that rise from their rest position in their housings on the seabed (the recesses) within the entrance channel and are subsequently lowered once more. These devices are activated when the tide level exceeds a preset safety level, currently defined as +110 cm above the altimetric reference point at Punta della Salute. The planned works also consist in the construction of complementary fixed structures to increase friction in the entrance channels in order to lessen the liveliness of the tidal currents, mitigating the levels of the most frequent ones.

The Lido entrance is the northernmost inlet in the lagoon, close to Venice, and is very wide and of varying depths. Therefore, at the design level, two mobile barriers of flap gates were planned, one on the Treporti channel to the north and one on the San Nicolò channel to the south, while a 9-ha artificial island was built in the center of the inlet to serve as an intermediate structure between the

← The Malamocco lagoon entrance.
→ Phases of the barrier lifting tests.

two rows of flap gates, housing all the technological buildings and systems for the operation of the two barriers. The north bank of the entrance, on the Treporti coastline, was enlarged and redesigned. A sheltered harbor was built, consisting of two protected basins, one on the seaward side and one on the lagoon side, which allow boats and emergency vehicles to moor and pass through the navigation lock when the floodgates are in operation, that is raised. Along the south side of the entrance, the existing wharf was extended, while on the outer side the curvilinear breakwater extending for about 1 km was completed to dampen the tidal currents entering the lagoon. All these works were preceded by major consolidation work on the seabed of the entrance.

• The Lido Treporti barrier is 420 m wide and 6 m deep. There are 21 gates, each 18.55 m long and 20 m wide, and 3.6 m deep. The gates are hinged to 7 concrete housing caissons, installed in the seabed, 36 m wide and 60 m long, with a height of 8.7 m. Each of these houses 3 gates. The 2 abutment caissons on either side of the barrier, named East Abutment and West Abutment, allow access to the tunnels that pass through the underwater structures. The barrier is also equipped with a navigation lock to allow the passage of boats—mainly pleasure craft and emergency vehicles—during the raising of the flap gates.

• The Lido San Nicolò barrier is 400 m wide and 12 m deep. This is the entrance used by cruise ships bound for the tourist port of Venice. It consists of 20 gates, each 26.6 m long and 20 m wide, and 4 m deep. The gates are hinged to 7 housing caissons which are 45.5 m wide and 60 m

→ The Lido entrance to the lagoon.

A
B
C

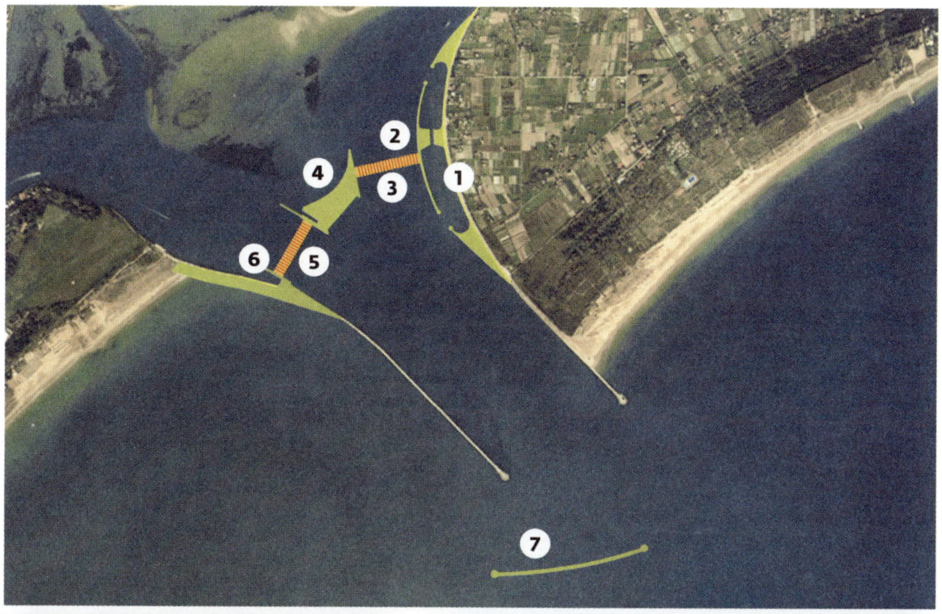

A. Lido Lagoon Entrance

Lido Treporti
width of barrier: 420 m
depth of entrance channel: 6 m
number of floodgates: 21
length of floodgate: 18.5 m
width of floodgate: 20 m
thickness of floodgate: 3.6 m
number of housing caissons: 7
length of caisson: 36 m
width of caisson: 60 m
height: 8.7 m

Lido San Nicolò
width of barrier: 420 m
depth of entrance channel: 12 m
number of floodgates: 20
length of floodgate: 26.6 m
width of floodgate: 20 m
thickness of floodgate: 4 m
number of housing caissons: 7
length of caisson: 60 m
width of caisson: 45.5 m
height: 11 m

1. shelter and navigation lock
2. abutment
3. row of floodgates
4. new island and abutments
5. row of floodgates
6. abutment
7. breakwater

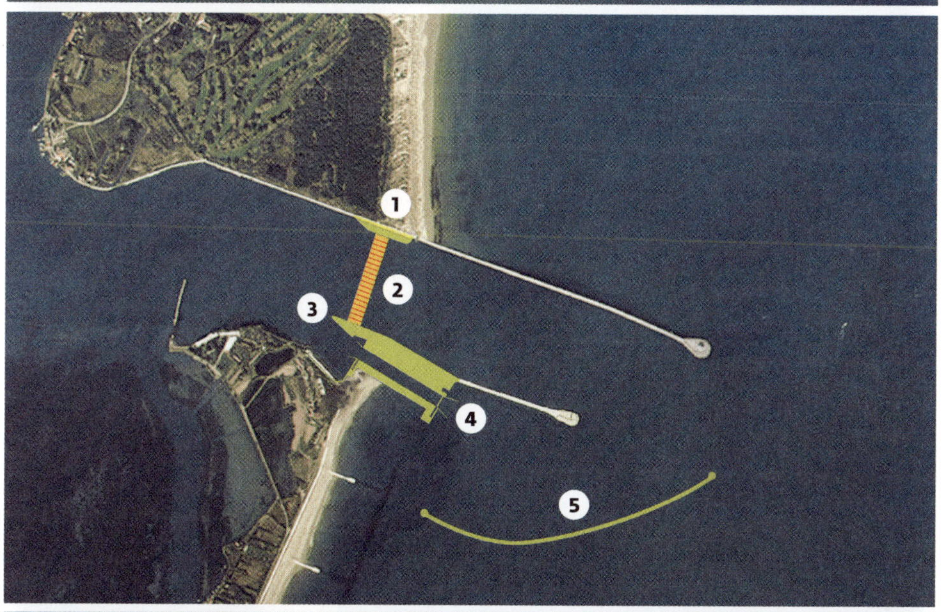

B. Malamocco Lagoon Entrance

width of barrier: 380 m
depth of entrance channel: 14 m
number of floodgates: 19
length of floodgate: 29.5 m
width of floodgate: 20 m
thickness of floodgate: 4.5 m
number of housing caissons: 7
length of caisson: 60 m
width of caisson: 48.3 m

height: 11.5 m

1. abutment
2. row of floodgates
3. abutment
4. lock
5. breakwater

C. Chioggia Lagoon Entrance

width of barrier: 360 m
depth of entrance channel: 14 m
number of floodgates: 18
length of floodgate: 27.3 m
width of floodgate: 20 m
thickness of floodgate: 5 m
number of housing caissons: 6
length of caisson: 46 m
width of caisson: 60 m

height: 11.5 m

1. refuge port and navigation lock
2. abutment
3. row of floodgates
4. abutment
5. breakwater

long, with a height of 11 m. In this case, 6 caissons were built for 3 gates each and 1 smaller caisson for 2 gates. The buildings constructed on the abutment caissons providing access to the tunnels are called the North and South Abutments.
In the specific case of the Lido, where two rows of flap gates have been built, the West Abutment buildings of the Lido Treporti barrier and the North Abutment building at Lido San Nicolò also provide access to the artificial island.

The Malamocco barrier is the deepest entrance to the lagoon at -14 m, and it is through here that ships bound for the commercial and industrial port of Marghera pass.
On the south bank of the entrance, on the island of Pellestrina, a navigation lock has been built to allow large ships to pass through during the raising of the gates. Outside the entrance there is a curvilinear breakwater that extends for about 1.28 m and serves both to dampen tidal currents and to create a basin of calm water to facilitate the maneuver of entering the navigation lock.
The Malamocco Canal is approximately 380 m wide and 14 m deep. This barrier has 19 flap gates, each 29.5 m long and 20 m wide, with a thickness of 4.5 m, hinged to 7 housing caissons.
The caissons are 48.3 m wide and 60 m long, with a height of 11.5 m. In this case, 5 caissons with 3 gates each and 2 smaller caissons with 2 gates were built. Buildings at the two abutment caissons on either side of the barrier, called North Abutment and South Abutment, allow access to the underwater tunnels within.

The Chioggia barrier consists of 18 mobile gates positioned along the entire 360-m gap between the two sides of the lagoon entrance. The wide channel, 11 m deep, is mainly used by fishing boats, pleasure craft and large vessels heading for the nearby Val da Rio docks. The final configuration of the entrance includes a breakwater on the seaward side (520 m long and 2.5 m high) to dampen the tidal current entering the lagoon.
The gates of this barrier are 20 m wide and 27.25 m long, with a thickness of 5 m. To house the 18 gates, 6 housing caissons and 2 abutments were built to connect the barrier to the banks of the entrance channel. Each caisson is 46 m long, 60 m wide, and 11.5 m high and was designed to house 3 gates. The buildings in correspondence with the abutment caissons, which allow access to the tunnels, are called North Abutment and South Abutment.
A large sheltered port was built on the north side to allow the passage of vessels when the gates are raised: it consists of two large basins (8 ha on the seaward side and 4 ha on the lagoon side), connected by a double navigation lock to cope with the heavy traffic of fishing boats, with peaks of over one hundred vehicles per hour. The two locks are 130 m long and 16 m deep.
The main buildings and installations necessary for the operation of the gates were built at the southern end of the barrier, as well as some areas for public use.

→ The Chioggia entrance during a phase of the works. The sheltered harbor was emptied of water to be used as a prefabrication area for the caissons • The Malamocco lagoon entrance.

The foundation caissons

The caissons housing the floodgates are prefabricated structures of reinforced concrete, built in temporary construction sites specifically set up for the purpose at the lagoon entrances. The caissons for the Lido San Nicolò and Malamocco barriers were erected on a levy built in the stretch of sea at the northern end of the Pellestrina coastline. The site was equipped to allow the prefabrication, handling, and launching of the structures.
← The prefabrication site at the Malamocco entrance.
→ Construction phases of the caissons for the Lido Treporti barrier.

The foundation caissons are the elements that form the base of the barriers to which the gates are attached. They are reinforced concrete structures built on land and then installed at the bottom of the entrance channel, within a specially dug and consolidated containment trench. The construction and installation of the foundation caissons (27 housing caissons and 8 abutment caissons) represented one of the greatest challenges in the construction of MOSE, due to the numerous critical aspects associated with the logistical, construction, organizational, and environmental factors. This required complex site management, specially designed executive systems and the use of the most advanced technologies in the field of maritime works. The result was certainly an excellent one, achieved through choices that were not only highly innovative but also bold as they differed from previous experience. Particularly significant in this respect are the tolerances of the underwater maritime works, which during the sinking and alignment of the structures were of the order of 1 cm. To achieve

these tolerances, it was necessary to develop *ad hoc* plant and working methods, and to train the personnel in a specific way.
The systems for operating the barriers and the tunnels for technical inspections were installed inside the caissons. The gates are connected to their housings by hinges that allow them to move.

The size and function of the caissons

The size and function of the caissons
The size of the caissons varies from barrier to barrier according to the depth of the seabed and therefore the size of the gates. The length, width, and height of the caissons for each barrier are shown below:

- *Lido Treporti:* : 60 × 36 × 8.7 m (weight approximately 13,000 t);
- *Lido San Nicolò:* 60 × 45.5 × 11 m (weight approximately 19,500 t);
- *Malamocco:* 60 × 48.3 × 11.5 m (approximate weight of 22,500 t)
- *Chioggia:* 60 × 46 × 11.5 m (weight approximately 20,400 t).

Aligned with each other, the housing caissons form the actual dam across the lagoon entrances, while the abutment caissons provide the interface between the housing caissons and the connection with the mainland, with descents within for equipment and workers. The most imposing structures of the barriers are the abutment structures. These are up to 28 m high at Malamocco (as high as an eight-story building) and have a surface area of 60 × 24 m.
The housing caissons are divided into compartments and internal rooms organized on several levels, and have a central part shaped according to the form of the gates. They are made of reinforced concrete and are equipped with sophisticated plant engineering

→ Detail of the construction of a caisson.

networks, which form the technological heart for the movement of the gates: these are located in the tunnel that crosses the caissons, divided into three tunnels (main, secondary, and drainage). Access is via the abutment buildings.

Caisson structure

The caissons have internal cells of varying dimensions, starting from 4 × 5 m in plan with heights of between 3 and 5 m, depending on their location. The cells intended to be filled with water or concrete are watertight by necessity; flooding of the cells, or filling them with concrete, was necessary when the caissons were lowered into the sea in order to counterbalance the forces of wave motion and improve the position of the caisson's center of gravity.
The central distribution tunnels were designed to be double and parallel, because each type of system—mechanical, electrical and hydraulic—was built with redundancy so as to be able to function even in the event of a breakdown.
All the cellular caissons have been prefabricated and assembled using the technique typical of immersed tunnels. Those for the mobile barriers required special attention in the choice of materials and corrosion protection systems, following evaluation of behavior under cyclic loads and the evolution of crack phenomena.
With slight differences within each site, each caisson was prepared in 9–12 months, with a minimum of 15 and a maximum of 23 casting phases. The total prefabrication time therefore meant that at least 2–3 caissons were built in parallel.
The construction of the precast caissons for the four barriers involved the use of

← A caisson ready to be launched at the Malamocco site. → Operational phases and interior of the caissons with the tunnels running through the entire barrier. For the sinking phase, the compartments forming the tunnels were sealed with watertight doors, which were removed as the caissons were joined together and aligned on the seabed; the connection and hydraulic seal is assured by a double system of joints • The omega joint. ↓ The construction of the caissons.

about 220,000 m^3 of concrete, with an average reinforcement rate of about 300 kg/m^3, but with peaks of up to 500 kg/m^3 for particularly stressed areas. The single-stage construction of the base slab involved the continuous casting of approximately 1,000 m^3 of concrete. Each company involved made its own choices in relation to the concrete mix, formwork, volumes, and casting phases, taking into account its own experience and site organization. However, the construction of the caissons for the various lagoon entrances was characterized by very similar methodologies.
In order to test the construction methodology and materials, a number of test castings of significant dimensions (mock-ups) representing a typical portion of the cellular structure of the caissons were successfully carried out on site.

Prefabrication of caissons

All the caissons for the various rows of flap gates were built in temporary sites close to the lagoon entrances in order to avoid interference between the sites and the surrounding area. This choice required a complex logistical, operational, and management organization, since the prefabrication areas, in preexisting waterside spaces, were far from the environmental or urban context of the coastline, so much so that the supply of materials and the transfer of work vehicles had to take place almost exclusively by sea.
Prior to the construction of the caissons, mock-ups were made of the threshold caissons, which were used to perform destructive and nondestructive tests to evaluate and calibrate certain materials and technologies that were to be used for the construction of the caissons for the Treporti barrier and their installation. The tests carried out on these mock-ups, which reproduced two cells of the Treporti threshold caisson on a 1:1 scale, made it possible to check the validity of the materials and technologies to be used and to calibrate the nondestructive tests to be performed on the structures. The mock-up, measuring 10.4 × 3.45 × 3.15 m, was built on the reclaimed land near Treporti on a foundation plate and on an adjustment plane entirely similar to the one that would later be used for the prefabrication of the sill and abutment caissons.
During prefabrication, two female elements of the hinge-connector assembly for each flap gate were also laid and arranged on the caissons, by embedding the box element with grout and then stringing the 10 perimeter bars.
The seats for the legs of the jack-up were made, aligned along the deck slab on the seaward and lagoon sides, with the construction of the temporary support saddles used during the sinking and launching of the caissons; jointing collars were made for the joints between the individual caissons and GINA-type joints were fitted before launching.
The sealing system was completed with an Omega joint, which was fitted within the installed caissons. The systems in the tunnels (mechanical systems, piping, breakers, temporary ventilation, and lighting systems useful for the post-sinking phases, etc.) were preassembled. Temporary turrets and buffers were built for the purposes of flotation, transportation, and launching of the caissons, and then removed.
As already mentioned, the caissons were built with different operations and construction site phases according to the

→ The entrance at Lido Treporti. Once the caissons were completed, the prefabrication site was flooded so as to return it to its function as a water basin and to allow the caissons to be floated and towed to the installation point by a special vessel. Here they were ballasted and sunk within the area of the channel bottom at the entrance, which had previously been excavated and prepared to house the barrier.

← ↑ The caissons for the Lido Treporti and Chioggia barriers were prefabricated in the seaward basin of the new sheltered harbor planned to allow the sheltering and passage of vessels through the navigation locks when the MOSE barriers are in operation. The area was temporarily waterproofed, emptied of water, and equipped to be used as a prefabrication yard at a working level of approximately -9 m at the Lido and -12 m at Chioggia. The housings for the Lido Treporti barrier were built within the seaward basin of the Punta Sabbioni harbor; similarly, the caissons for the Chioggia barrier were built in the seaward basin of the Ca' Roman harbor.
↓ Section of the seaward basin emptied of water.

The MOSE mobile barriers consist of three main elements: floodgates, caissons, and hinges. The gates are inserted into their housings—that is, the caissons positioned in the bottom of the channels of the lagoon entrances and aligned with each other. The array of submerged housing caissons is connected to land by means of abutment caissons located at either end of the barrier. The gates are attached to the caissons by means of hinges. The caissons also house the plant and piping for operation, which are connected to the gates via the hinges.

← ↑ Setting up of the caissons inside the prefabrication area; before sinking, the female parts of the hinges—the yellow elements—were installed in the caissons.
↓ A caisson with the female hinge element and the compartments corresponding to the double row of tunnels.

peculiarities of the various barriers.
The housing and abutment caissons were assembled within the seaward basin of the Punta Sabbioni (Lido Treporti) shelter harbor. For this project, the basin was waterproofed and emptied of water, creating a working surface at a height of -8.7 m, which was equipped as a worksite area where the abutment structures were built in addition to the housings. Once all 9 caissons had been built (2 abutments and 7 housings), the basin was reflooded, and one by one they were towed along the channel to the point of installation with a special boat, where they were sunk and ballasted in the excavation in the seabed.
At the Malamocco entrance, on the temporary berm created at the northern end of the Pellestrina coastline, both the Malamocco caissons and those of the Lido San Nicolò barrier were built.
The prefabrication area and the Syncrolift area were also set up here.
The sinking and flotation of the caissons took place in Malamocco using the Syncrolift.
Once afloat, the caissons were hooked up and towed to the barriers at Malamocco and San Nicolò for subsequent sinking.
The prefabrication area was equipped with hundreds of concrete "pillars," arranged in parallel rows perfectly aligned with each other, which supported each caisson at a height of more than 2 m to allow it to be lifted and moved into the Syncrolift Area.
The Syncrolift is a huge lift capable of lifting a weight of about 30,000 t, enabling the caissons of Lido San Nicolò to be carried from the Malamocco site.
The launching platform was raised and lowered by special steel cables of a diameter of 4.5 cm controlled by electric winches.

The caissons were built on top of a series of pillars and, once completed, were lifted by means of a hydraulic trolley system and transported on rails to the Syncrolift launching platform.
To launch the caissons —that is, to sink them in water—a special platform was planned which, by means of a system of winches, made it possible to immerse the caissons in water, tow them to the final position in the lagoon entrance channel, and finally ballast them on the seabed.

← The caissons under construction in the prefabrication area at the Malamocco inlet and the construction of the pillars on which the caissons were built.
→ Detail of the transport system for transferring the caissons to the Syncrolift.

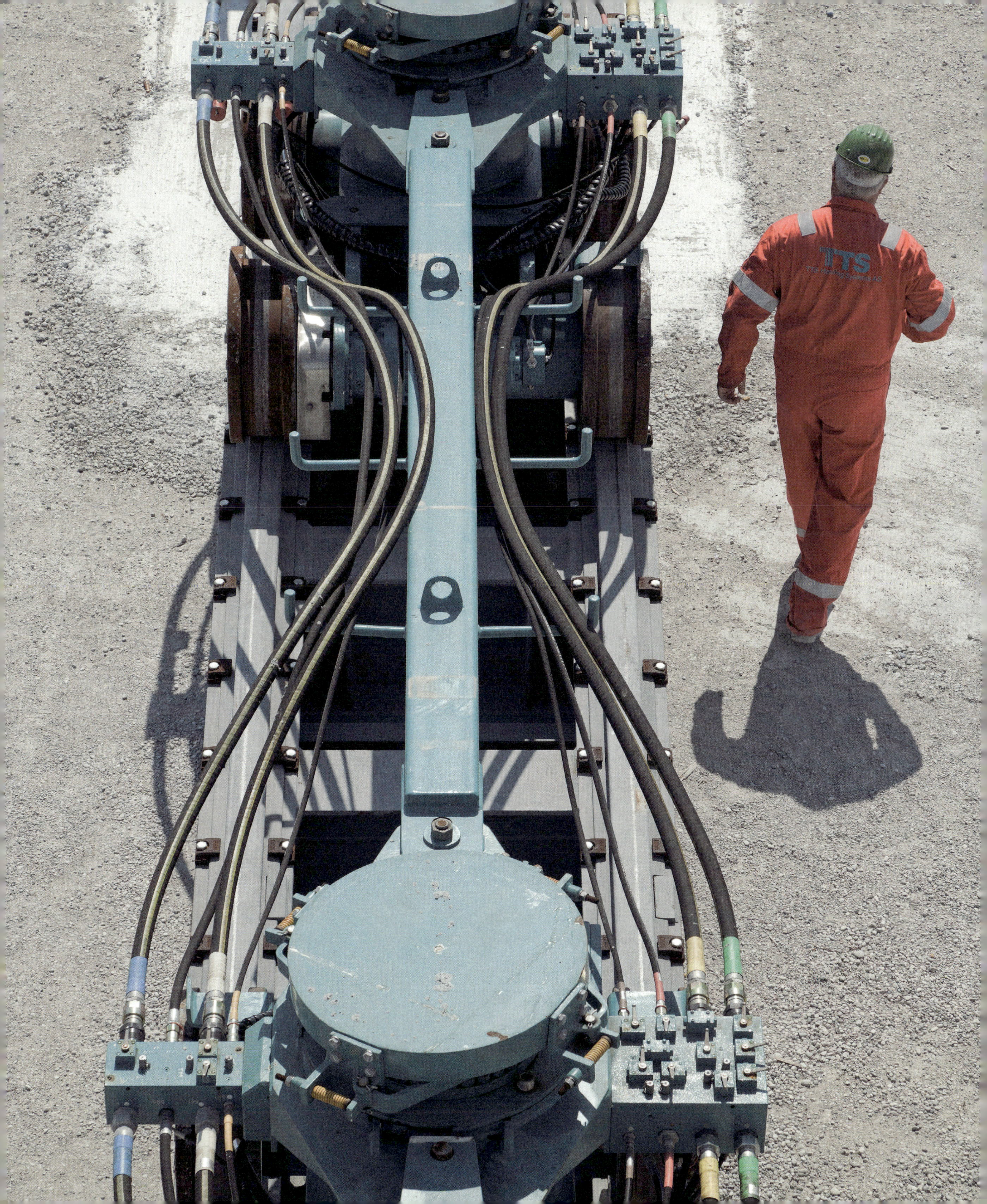
TTS

The platform deck consisted of a frame of 13 main steel girders, 57 m long and 5 m high, complemented by a series of smaller secondary girders and a floor of high-strength wooden planks. The system was patented by Rolls-Royce Naval Marine, Inc. (Annapolis, US) and the result is the largest platform of its kind in the world. The site was set up on the vast berm realized and equipped on the south bank of the Malamocco entrance, which is currently being dismantled.
The movement of the caissons required a special system of rails and hydraulic steel bogies, developed by TTS Handling Systems AS (Drøbak, Norway), which allowed each element to be moved as if it were a railway carriage. Due to the size of the housing caissons, 70 large bogies were required for each, divided over 6 rows of rails, following a precise scheme that predetermined the exact position of each according to the progress of the main structural parts. The rails and bogies were inserted under the caissons, in the space created by the pillars that kept them raised off the ground.
Thus, sliding on almost 300 wheels (4 for each trolley), the caisson reached the launching platform within a few hours and was perfectly positioned in the center of the large Syncrolift deck.
The next step was carried out with equal caution and involved lowering the platform into the sea at a speed of 10 cm per minute until the caisson was floating, secured and controlled with large cables.
In Chioggia, as at the Lido entrance, all 8 caissons were built at the sheltered port of Ca' Roman.
The construction site was set up within the seaward basin, which was temporarily isolated from the entrance

↑ The winch system for moving the Syncrolift caisson launching platform • A trolley positioned under a caisson for moving.
→ The rail system in the Malamocco prefabrication area • Moving a caisson • A phase of setting up the Syncrolift launching platform.

channel, waterproofed, and emptied of water, as was the case for the barrier at Lido Nord.
In this way, a large construction area was obtained at a depth of 12 m below sea level, which was kept dry for the entire duration of the work by means of a sophisticated drainage system.
Once the construction of all caissons was completed, complete with plant systems and hinges, the dock was reflooded and opened to allow the structures to float out.

Launch and sinking of the caissons

The transportation and installation of the barrier caissons required a large number of complex operations, supply logistics and the organization of the installation.
In order to highlight the difficulties involved, the transportation and installation of the caissons from the Malamocco barrier to the Lido San Nicolò barrier is illustrated below.
The caissons in the prefabrication area were moved by a system of rails and bogies to the Syncrolift launching platform.
Towed by special tugs and the pontoon—which in the case of the Lido San Nicolò barrier involved sailing a few kilometers off the coast where the seabed is sufficiently deep—the convoy entered the entrance channel and reached the planned position.
Here it was moored while awaiting the right moment for installation. The experts at the MOSE Control Center provided constant up-to-date weather and sea forecasts, which were useful for determining the "go-ahead" moment for the various activities well in advance.
Once the "go-ahead" for installation had been given, the caisson was lowered to the seabed and positioned within the

SWL

← An intermediate stage of the Syncrolift descent with the caisson ready to be transported floating and installed in the barrier area.
↑ Phases in the launching of a caisson at the Malamocco entrance. The caisson is transferred to the Syncrolift launching platform for the subsequent lowering into the water. On the caisson, in addition to the female hinges, some provisional accessories in yellow are visible, prepared for the launch: the mooring cable protection structures, the towing bollards, the towing stands for attaching to the towing winches, and the towers for temporary access to the caissons during the installation phases.

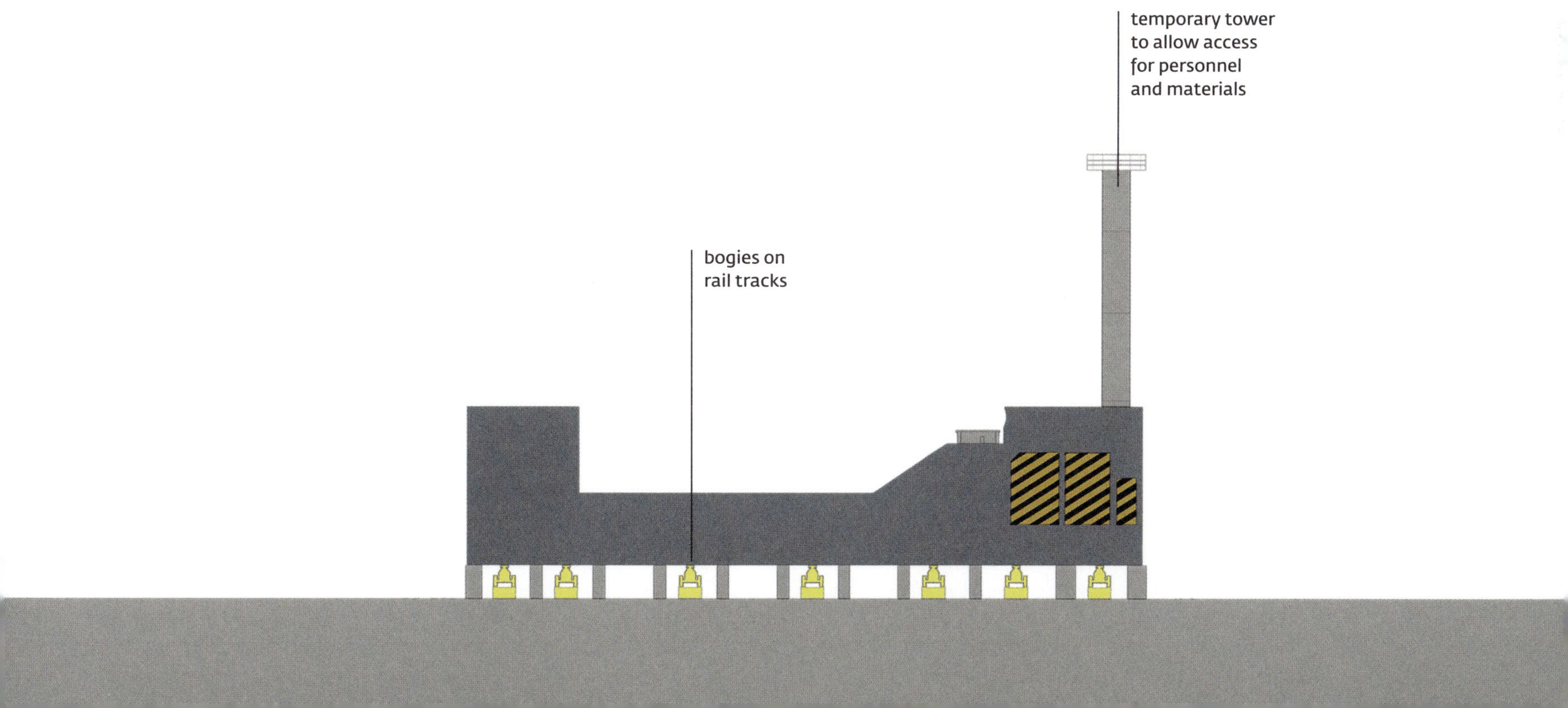

special excavation previously made along the entire length of the Lido barrier. Within this excavation, the caissons were placed side by side and aligned with each other with very little margin for error, as if they were components of a precision mechanism, despite the size of the objects. Since they already arrived with the necessary amount of ballast, all that was needed to lower them to the bottom was the controlled loosening of the cables that kept them afloat and bound to the installation pontoon, progressively accompanying their descent. All this, however, was done in the absence of wind and waves and on the day of minimum tidal excursion (during the lunar quadrature phase), with only a very slight current. Proceeding at a rate of 40 cm per minute, the caisson reached its underwater site after about one hour.
At this point, the final and most important adjustments were made, which were used to record both the vertical alignment of the caisson using powerful jacks placed at the four corners of the caisson base, and the horizontal alignment using specially designed instruments and devices.
Finally, the launching of the caissons at Lido Treporti was carried out using another construction site vehicle called the "Catamaran," consisting of two barges connected by special spacer beams and equipped with all the necessary equipment for the transport and laying of the caissons in the barrier, including appropriate pushers.

Barrier abutment buildings

The abutment buildings are located on the surface above the abutment caissons and contain all the control systems necessary for raising the gates, as well as for air conditioning the tunnels.

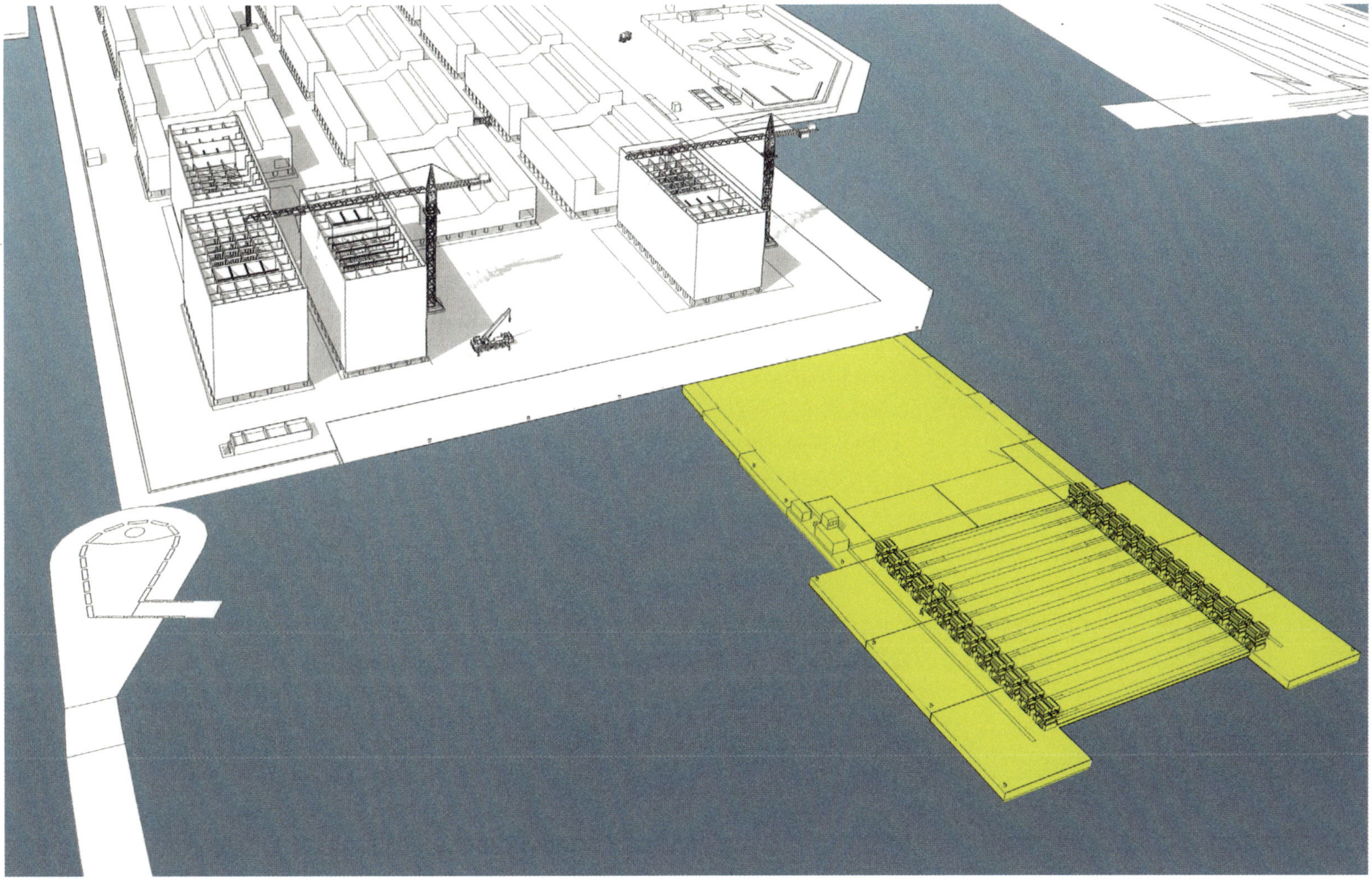

← Stages in the installation of a caisson.
↑ The prefabrication area at the Malamocco entrance with the caissons completed and ready for launching. The Syncrolift appears in yellow.
↓ A caisson positioned on the launching platform and ready to be submerged in water, transported, and installed in the seabed.

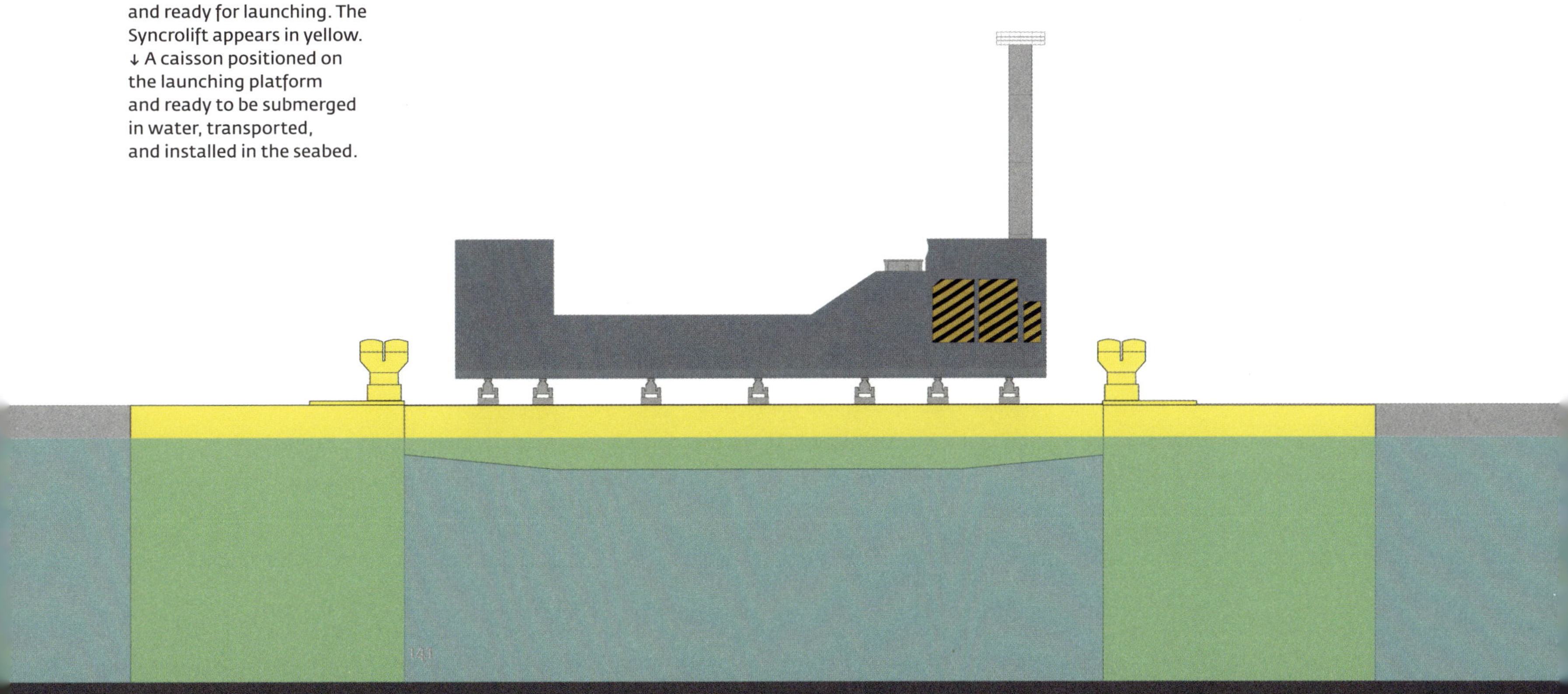

The abutment buildings and the technical area behind, where the main plant rooms are located (compressors, generators, electrical building), are protected on the water side by breakwater walls.
The buildings are on several levels: as a general rule, the lower floor, which is completely underground, lies at 0.5 m above sea level. It houses the electrical switchboard room for the cathodic protection system of the abutment Larssen sheet piling, the fire-fighting cylinders, the HVAC rooms, the black water collection tanks, the cable ways (electrical walkways) from the tunnels of the barrier to the switchboard rooms on the upper floor, and the stairwells connecting the basement with the upper floor. The two tunnels connecting the barrier to the plant buildings also run along the sides of the building at this height, so that all the pipes and cable routes from the tunnels descend into the floor at a height of +0.5 m to the plant shafts, which in turn run to the barrier tunnels.
The ground floor of the control buildings at an elevation of 3.5 m houses the rooms containing the electrical and instrumentation systems for powering and controlling the systems serving the gates, such as the electrical rooms, UPS battery rooms, etc.
It also houses the guard rooms, rooms for service personnel, offices, meeting rooms, infirmary, changing rooms, and toilets. At the same level, there are also the corridors providing access for service vehicles to the shafts, through which the necessary spare parts can be lowered down to the tunnels via two monorail hoists.
On the upper floor, usually at an elevation of 7.6 m, there are the control room, the telecommunications room, and the fire-fighting room.
Finally, at the various lagoon entrances, external arrangements have been provided for the appropriate architectural inclusion of the buildings into the setting.

→ The new island at the center of the Lido lagoon entrance connecting the two barriers and housing the area with the technical buildings for operation and control.

Technological buildings

At each lagoon entrance, next to the abutment buildings, there are the technological structures for the systems serving the barriers from which they branch off, and the shafts going down to the tunnels and the connector rooms. The main technological hub of the system is on the artificial island of the Lido barrier (Isola Novissima), which feeds both the Lido Treporti and Lido San Nicolò barriers. The artificial island has a surface area of approximately 90,000 m^2, while the built-up area for the plant has a surface area of approximately 16,000 m^2, with a volume of approximately 800,000 m^3. The technological center is completely underground and the various technical buildings are interconnected by tunnels. The most important buildings in the barrier areas are the electrical building, the generator building, and the compressor building:

• *the electrical building* is mainly underground and has a cable duct underneath to distribute the cables to the utilities. Inside, two isolated rooms house the medium voltage switchboards, the MV/LV transformers, the low voltage switchboards, the transformer cooling fans, the cooling and ventilation units for the electrical rooms, the fire prevention room, etc.;

• *the generator building* is essentially underground, with smaller bodies above ground and a deeper offshoot where the waste oil tanks are located. This building houses the emergency electric generators, with their engine and alternator, suitably enclosed for soundproofing, the tanks (alongside the machine, the daily diesel oil tank, the gravity collection tank for the diesel oil overflow pipes, and the spent oil loading and unloading tank), the electric panels for the generators, the power panels for the service equipment, the equipment for ventilating the rooms and cooling the control rooms, the cooling air fans, the filters, and silencers;

• *the compressor building* is for the main part underground, except for the filtration and noise reduction units, which protrude a couple of meters above ground level. The building houses the air compressors for lifting the flap gates, the medium-voltage electrical control panels and the compressor control panel, the electrical panels for the HVAC, hoist, light, sockets and ventilation of the compressors in the relative rooms, the instrument panels and the panel for paralleling the units, the cooling and ventilation units for the electrical rooms, and the compressor rooms.

← The prefabrication area at Malamocco with the Syncrolift launching platform and caissons.
↓ The caisson prefabrication area at Lido Treporti before and after flooding for the launch of the structures.

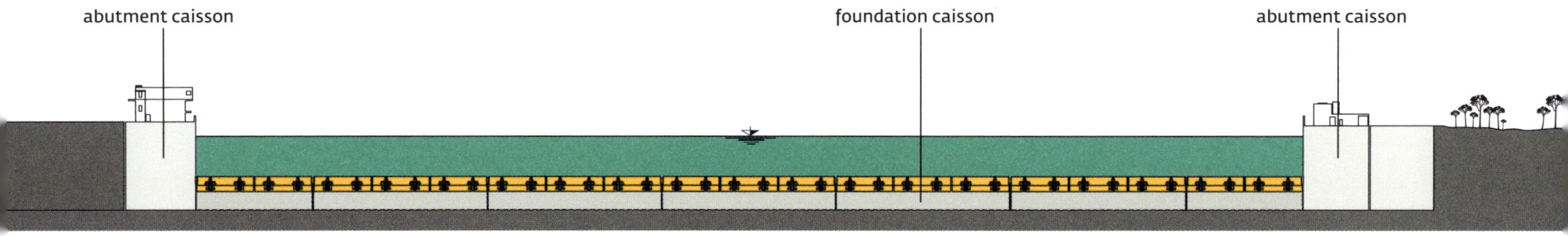

Malamocco lagoon entrance

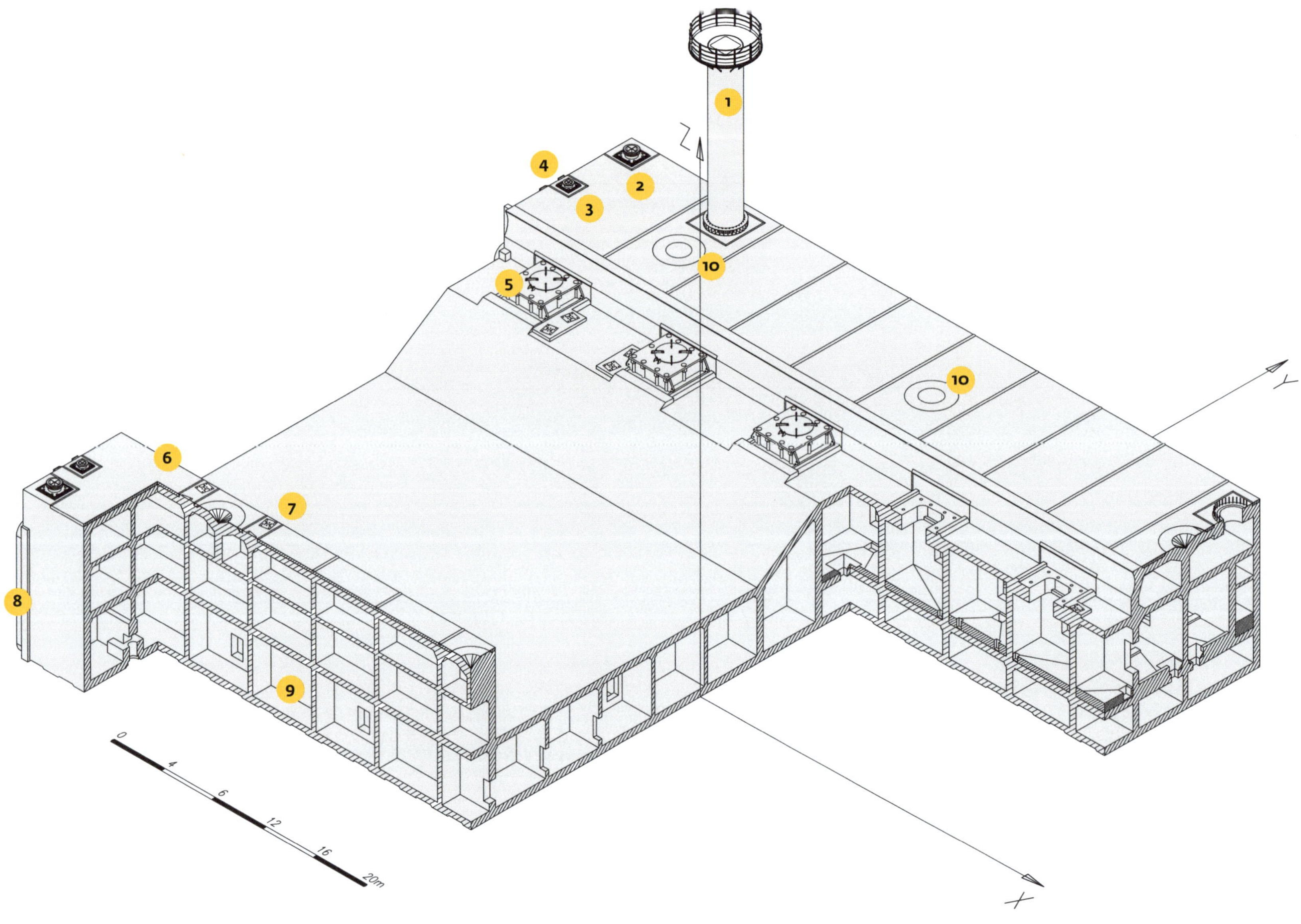

Axonometric perspective of a caisson and section of the barrier with the row of sill and abutment caissons. Each barrier consists of two abutment caissons, which connect it to the land, and a series of submerged sill caissons, which house the gates. The abutment caissons are located at both ends of the barrier, connect the underwater array to the land, and allow entry and exit to the technical rooms inside the tunnels.

↑ **foundation caisson**
1. Access and alignment towers (2)
2. Double bollard, 300t (4)
3. Pulling bollard, 100t (4)
4. Male coupling unit (2)
5. Female connector unit (6)
6. Protective screed for roofing slabs
7. Base plates for lifting frame installation barge coupling (16)
8. Joint-bearing collar
9. Balancing ballast
10. Mounting plates for jack-up supports (6)

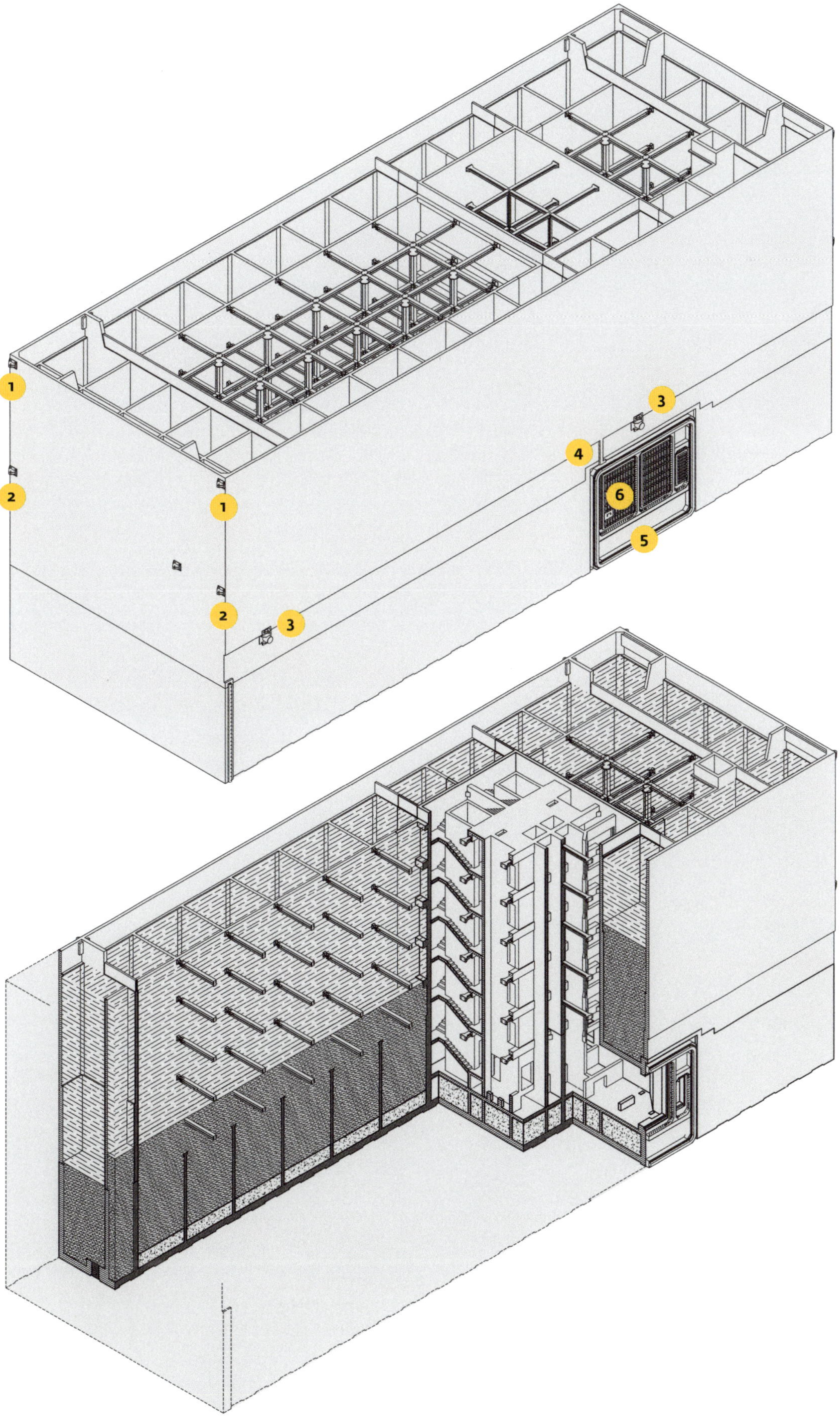

↑ abutment caisson
1. 100t lateral docking coupling
2. 215t lateral towing/docking coupling
3. Coupling for pulling jacks + coupling guide (male)
4. Beam-supporting corbel
5. Joint-bearing collar
6. Metal buffers

The locks

The navigation locks are located at Lido Treporti, Malamocco, and Chioggia and allow the transit of vessels when the barriers are raised.

The Lido Treporti lock

A navigation lock has been built at the mouth of Lido Treporti along the Cavallino bank to the east of the barrier, with an internal length between the gates of 70 m, a width of 18 m, and a minimum depth of 5 m. The dimensions have been established for small boats and fast passenger craft, particularly hydrofoils and catamarans. Longitudinally, the basin is built with combined walls of metal sheet piling, anchored at the top by steel tie rods. Radial gates have been installed for closing and opening, which can be operated without waiting for the equalization of levels between the lagoon and the sea, with the difference in level supported even when there are currents at play, thus reducing the cycle time for each lock-through. This makes it possible to keep the gates open during the safety period of about 30 minutes before raising the flap gates. Each lock gate appears from above as a circular sector described by a radius of 9.3 m at an angle of 90°.

← The Chioggia entrance to the lagoon. On the left, the sheltered harbor with the two basins and the double navigation basin.
→ The Malamocco entrance to the lagoon • The Chioggia lock basin • Entrance to the Malamocco lock.

It consists of a strong metal frame covered along the circular arc by a continuous sheet metal shell. The supports consist of two hinges: the upper one supports the weight of the structure while both are able to take up the sum of the horizontal forces. The device for the door opening and closing maneuver involves the use of an electric motor, which transmits the rotary motion through a pinion-rack coupling.
The opening maneuver of the lock gates is carried out in two stages. Firstly, each gate is rotated only a few degrees to enable communication between the level inside the lock and the level outside the lock, with appropriate baffles between the abutment and the gate. Later, when the difference in level has been reduced to 10÷20 cm, the rotation of the door is completed in one go.

The Malamocco lock

The lock basin has been designed to allow the passage of large ships during the operation of the barriers and has internal dimensions of approximately 384.5 × 50.4 m; it is bounded by longitudinal walls with an extrados of 3.5 m above sea level and, at the transverse ends, from the lock gates and from the structures to house them, with an extrados of 2.7 m above sea level. The level of the bottom of the basin is 13.5 m below sea level.
The side walls are made of a double face of Larssen sheet piling, reinforced at the top. The lock basin is terminated at the end by two sliding steel gates and consists of embankments bordered by combined steel tensioned walls, each with a reinforced concrete crowning beam at the top. The lock gates rest on a friction-reducing "hydrofoot" system and are moved by means of a cable and winch system. Each of the two gates consists of an open steel lattice structure, with

→ The sheltered harbor at Lido Treporti with the two basins and the lock.

plating on the lagoon side. Each gate is installed inside a housing known as the "large recess" and slides over a sill arranged longitudinally to the axis of the gate; the so-called "small recess" is located at the arrival position of the gate. During the storm surge of February 5–7, 2015, the seaward lock gate was damaged and consequently redesigned. New marine studies and physical model tests have been carried out in cooperation with Deltares and Maritime Research Institute Netherlands (MARIN), and work on the new seaward gate and work on the lagoon-side gate is still ongoing.

The Chioggia locks

The two basins of Chioggia, for the passage of small boats and fishing vessels, are located along the northern side of the lagoon entrance. The lock basins measure approximately 130 × 16 m, are delimited by longitudinal walls and, at the narrow ends, by the lock gates and structures used to house them. The bottom of the basins is 5 m below sea level. As in the other chambers, the side walls of each chamber are made of a double wall of Larssen sheet piling, which is tensioned at the top, and the basins are closed by vertical steel gates. Each of these is installed within a housing structure which provides stops and recovery when open, leaving the mouth of the basins completely clear. There are 8 gates, one pair for each entrance to the lock, and they regulate the flow of water and levelling operations by gradual opening. The four pairs of gates are identical not only in the geometry of the mouth of the basins, but also in the volume of water they support and their buoyancy stability.
Each local gate appears from above as a circular sector described by a radius of approximately 8.3 m at an angle of 90°. The door consists of a metal frame covered along the circular arc by a continuous sheet metal plating; the supports consist of two hinges capable of taking up the resultant horizontal forces while the lower one supports the weight of the structure. The device for opening and closing the gate involves the use of a hydraulic cylinder that transmits pulling and pushing actions on the gate frame, which translate into rotational motion around its vertical pivot.

← The Malamocco entrance with the navigation lock. In the foreground, the arrival of the first floodgates to the area used to prefabricate the caissons, which, following their installation, were used to store the floodgates before they were laid in position.
↓ The navigation lock at the Malamocco entrance.

Hinges

← A floodgate before painting, with the male hinge elements. → A floodgate ready to be laid, with the male elements of the hinge installed • Processing stages of the female elements of the hinges ↓ Steps in launching a caisson.

The hinges, one of the most technologically innovative components of the entire MOSE structure, bind the floodgates to the housings on the caissons and enable them to move. They consist of a male element (3 m high, weight 10 t) connected to the gates, a female element (1.15 m high, weight 24 t) connected to the foundation caissons and a coupling unit that joins the male and female elements. A total of 156 hinges are planned for the 4 barriers, plus 8 spares.
The hinges have been manufactured using advanced systems for processing and welding steel because they have to withstand the stresses produced by the tidal stream and the wave motion on the gates. The calculated loads are several hundred metric tons, much higher than those for similar components in the offshore industry, and they must survive underwater in the highly aggressive marine environment. Prior to serial production of the hinges, a 1:1 scale prototype was built and subjected to four million tests.
The coupling and tensioning assembly is composed of various subassemblies that enable the main function of

GIUL

coupling and uncoupling the hinges of the flap gate; as well as establishing the structural constraint of the male–female connection, the assembly must also allow the passage of air to and from the flap gate to raise it, of the washing system for removing debris, of the electrical connection that powers the instruments positioned on the flap gate and the mechanical inclinometer contained within. The coupling and tensioning unit is operated by means of a hydraulic control unit located in the tunnel, outside the connector room: its connection to the various actuators and sensors is via a watertight multi-coupling connection plate. Operation is automatic and managed by software.
The female element is made up of welded steel plates that compose a box structure in the form of a parallelepiped, inside which the truncated cone housing the male unit is located. The female element consists of a contrast plate, a perimeter seal, movable sealing covers of the anchorage bars, elements for monitoring the cathodic protection, and strain gauge washers for tension monitoring. The fixing of the female element to the foundation is ensured by 10 post-tensioned anchorage bars, arranged along the perimeter and housed in special tubes present both in the metalwork and in the concrete structures. The box is subdivided internally by 6 vertical reinforcing plates, which connect the central truncated cone seat with the cylindrical ones for the bars and create as many compartments communicating with each other through the holes and drainage grooves in the plates. Inside the plates are the venting tubes for the fixing grout, the passages for the connection cables to the reference electrodes and the junction-boxes where these cables are laid. The space between the male conical element and the female truncated conical seat is called the lateral gap; the contact areas with the seals of the male are coated with a material with high resistance to corrosion in seawater (Inconel 625).
As set out in the project, all surfaces of the female unit in contact with seawater are painted. Externally to each female element, 4 sacrificial zinc anodes are installed, mounted on special supports and connected by shielded electrical cable to the junction-box positioned inside the box element. Their function was to protect the hinges from corrosion during the transition phase, namely during the period between the laying of the hinges and the laying of the gates.
It is the male element connected to the flap gate that enables the same to be linked to the female element. This is made from very thick welded sheet metal, mostly 120 mm thick. The central plates are approximately 2,500 mm high and are welded to a circular lower plate with a diameter of 1,250 mm. Pairs of ribs 80 mm thick and almost triangular in shape stiffen the main plates in the orthogonal plane, both above the upper plate and between the two discs.
In the lower part of the truncated cone volume, the inner side of the plates is machined in such a way as to obtain the seat of the saddle for housing the head of the coupling rod. The passage of air/water from the lower to the upper disc takes place next to two tubes arranged planimetrically on the orthogonal axis to that of maximum sea/lagoon engagement. The two outlets on the upper disc allow connection to the same number of manifolds by means of flanged connections, from which, again by means of flanged connections,

NOLEGGIO

← ↑ The male hinge elements and, below, the female hinge elements. The connection between the two elements ensures that the sluice gates are secured to their housings. The female elements are yellow parallelepipeds which are hollow on the inside to allow the insertion of the conical male element connected to the floodgates; they are then secured by the so-called coupling group which clamps the interconnection between the two elements. The females were installed on the caissons before launching and closed with special "plugs," which were removed for the laying of the floodgates, which ends with the touch down, namely the connection between the two elements of the hinge.

flexible hoses depart to convey the air to the piping integral with the flap gate. On the main axis of the element, in the space between the two vertical plates, there are housings for the electrical connector and the mechanical inclinometer.
The conical coupling characterizing the upper flange of the male unit reaches a stop on the (female) seat so as to transfer the preload action introduced by the stem and its traction mechanism.

Tensioners

The coupling unit consists of the following elements: coupling rod, segregation valve, guide sleeve, hydraulic mechanism for axial movement of the rod, hydraulic mechanism for the application of the pre-tensioning force, rod rotation device, crossbar for the pre-tensioning cylinders and mechanical locking device. The coupling rod, weighing approximately 860 kg, trasnmits the planned preload to the male connector and can basically assume two positions: the first, in operation, when the hammer head engages on the two coupling saddles, fixed to the male connector; the second, at rest, during insertion and subsequent disengagement of the male connector inside the seat of the female element, when the hammer head lies below the segregation valve inside the watertight housing.
The maneuver is carried out by means of a hydraulic control unit complete with control panel interfaced to the general control system. This is transportable and is positioned outside the connector room and connected to the through-wall flange, located at the side of the watertight inlet door, by means of a multicoupling flange.
The control unit, weighing approximately 500 kg, is mounted on a mobile unit which is moved by the operator along the service tunnel until it arrives near the entrance door to the connector room, in front of the through-wall flange. The operator effects the electrical and hydraulic connections outside the room and, after being cleared to proceed by the remote tensioner control system, completes the unhooking and hooking of the unit, remaining safely outside the room for the duration of the operational sequence.

← The male elements of the hinge.

The floodgates

At the heart of MOSE are the rows of retractable flap gates, which are independent of each other and remain invisible when inactive and full of water, laid in housings located at the bottom of the channels of the lagoon entrances. The barriers come into operation to interrupt the entrance of the sea into the lagoon temporarily when a high tide is expected and which might cause flooding of the area. They remain operational only for the duration of the event and when the same tide level is reached in the lagoon and sea, they are filled with water again and return to their seat.
The flap gates are modular steel structures, equipped with 4 fenders and rubber flaps; their internal and external surfaces are treated with special anticorrosion and antifouling paints, devoid of biocides. The purpose of the fenders is to dampen the impact of the flap gate on the caisson at the end of the lowering phase and to support the flap gate when it is at rest. The rubber flaps are arranged along the edges of the upper plating on the seaward and lagoon sides and are designed to reduce the air gap to limit the deposition of sediment in the

← A barrier in operation. → A floodgate before being painted, in the storage area where the male hinge elements were assembled • The inside of a floodgate with the (red) pipes for passage of compressed air • Transportation of a floodgate.

compartment below the flap gate when it is lying in the caisson and to reduce the air gap on the seaward side when the barrier is raised, limiting the flow of water generated by the hydrostatic head between sea and lagoon. They also provide an operating margin during the removal and installation of the flap gate, thanks to their elasticity, to avoid damage to the corners of both the flap gate and the caisson in the event of inaccurate maneuvering.
As mentioned in the description of the barrier configuration, 2 mobile flap gates were installed at the Lido entrance (Lido Tre Porti and Lido San Nicolò), 1 at Malamocco, and 1 at Chioggia, for a total of 78 plus 2 reserve flap gates for each row. As they are proportional to the depth of the inlet channel where they are installed, they have different widths. The total length of the 78 gates defending Venice and its lagoon is 1.56 km. To summarize, we will repeat the dimensional characteristics of the barriers:

• *Lido Treporti barrier:* 21 flap gates plus 2 reserve gates; channel depth 6 m; width 20 m, length 18.55 m, thickness 3.6 m; weight 168 t;
• *Lido San Nicolò barrier:* 20 flap gates plus 2 reserve gates; channel depth 12 m; width 20 m, length 26.65 m, thickness 4 m; weight 282 t;
• *Malamocco barrier:* 19 flap gates plus 2 reserve gates; channel depth 14 m; width 20 m, length 29.5 m, thickness 4.5 m; weight 330 t;
• *Chioggia barrier:* 18 flap gates plus 2 reserve gates; channel depth 12 m; width 20 m, length 27.25 m, thickness 5 m; weight 289 t.
Each flap gate at the San Nicolò, Malamocco, and Chioggia entrances

→ The Malamocco entrance with the barrier in operation.

is made of welded steelwork with a ship-like structure and has absolute pressure transmitters, electronic inclinometers with optical reader, inclinometers with magnetoresistive reader, and mechanical inclinometers installed.

Construction of the floodgates

The first 21 floodgates for the Lido Nord barrier were manufactured by Cimolai of Porcia (Pordenone) and installed in the Lido Treporti canal. The production of the other 57 floodgates (plus a further 6 reserve gates, 2 for each barrier) was carried out by the Brodosplit company of Split (Croatia).
The gates were transported by sea on pontoons towed by tugs. The first 21 sluice gates for the Lido Treporti barrier arrived from Monfalcone, while the other 57 gates for the Malamocco, Chioggia, and Lido San Nicolò barriers arrived from Split on special pontoons carrying four sluice gates at a time.
The laying of the gates was carried out at the Lido Treporti barrier in 2013 with the basic version of the launching gantry; at the Malamocco barrier between February and July 2017 and at the Chioggia barrier between September 2017 and January 2018, with the CV.02 version of the launching gantry; finally, at the Lido San Nicolò barrier between September 2018 and January 2019 with the Mose I jack-up.
The vessels used for laying will also be used for maintenance activities.

↑ Transportation of floodgates.
↓ Floodgates in the storage area at the Malamocco entrance before installation. In the background the gantry used for installation.

Principle of floodgate operation and process sequences

The floodgate operation phases can be summarized as follows: preliminary operations with drying of the lines, emergence phase of the floodgates, lifting to working angle, tide-tracking, lowering and return to recess, and finally emptying of the systems.
During the line drying, carried out in groups of 2–3 gates at the same time, one compressor is active for each sector. The "sectors" are the subdivisions into independent, fire-protected compartments of the floodgate lifting systems.
In the first phase, the floodgates are initially completely filled with water and laid horizontally on the bottom in the recesses: the processing plant, controlled by the Process Control System, starts to introduce compressed air which, due to the conformation of the pipes inside the gate, is released at the opposite end to the hinges. During the movement, an air bubble forms inside, which is concentrated at the end opposite the hinges to maximize the moment of Archimedes' thrust; when this exceeds the moment given by the weight force, the floodgates detach from the bottom. Replaced by air, the water is expelled through openings. During the surfacing phase, air is introduced into the floodgate, with flow rate regulation by the loading valve. The phase ends when the emersion angle is reached.
The lifting phase to the working angle starts when all the units have finished surfacing and involves all the gates at the same time. Air is introduced into the floodgates with flow rate adjustment by the loading valve until the working angle is reached. The working angles in the project for the various barriers are: Treporti 40°, San Nicolò and Malamocco 45°, Chioggia 42.5°.
During the tide-tracking phase, the floodgate is maintained by the control system at the working angle by means of letting air in or out for the entire period of closure of the barrier, maintaining a difference in water level between the sea and the lagoon. The feedback to the valves is based on the angular position: if the angle is outside the expected range (working angle +/- tolerance) air is injected or discharged to return the element to the correct position.
If the maximum set angle is exceeded, the Emergency Shut Down system intervenes.
Once the lowering command has been given, the system empties the sluice gate of air by opening the discharge valves at variable degrees of opening in order to avoid too abrupt a descent: first there is a lowering from the working position to the surfacing position, then the inclination decreases from the surfacing angle to the minimum floating angle after which, continuing the air discharge, the sluice gate sinks definitively and completes its descent into the recess.
Once recessed, the discharge valves are closed for a few minutes to separate water and residual air, and then opened again to let out all the air until the lines discharge only water for a sufficient time; at this point the floodgates are again at rest and secure. The water reenters the floodgates through the same openings from which it had escaped.

FAGIOLI
FAGIOLI

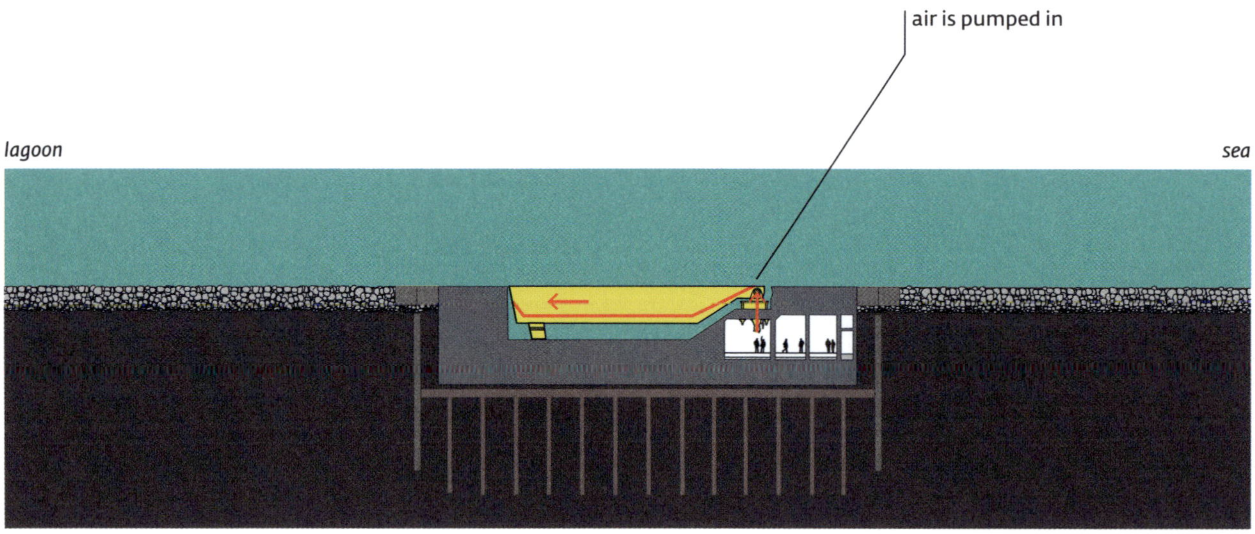

1. In normal tidal conditions, the floodgates lie flat in their housings, full of water

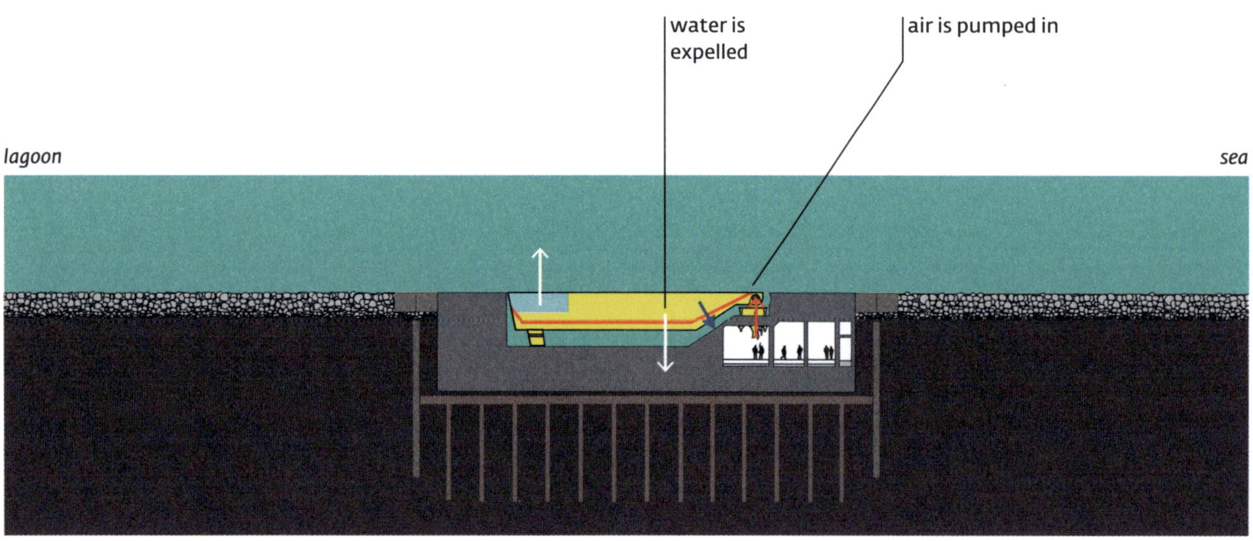

2. When a high tide above the warning threshold is forecast, they are emptied of water by means of compressed air. by means of compressed air. A system of compressors, located in the technical rooms of the abutment, enables the transmission of compressed air to the floodgates; passing through pipes in the underwater tunnels and connected to the hinges, the air reaches the interior of the floodgates, expelling the water that emerges through special openings on the back of the gates

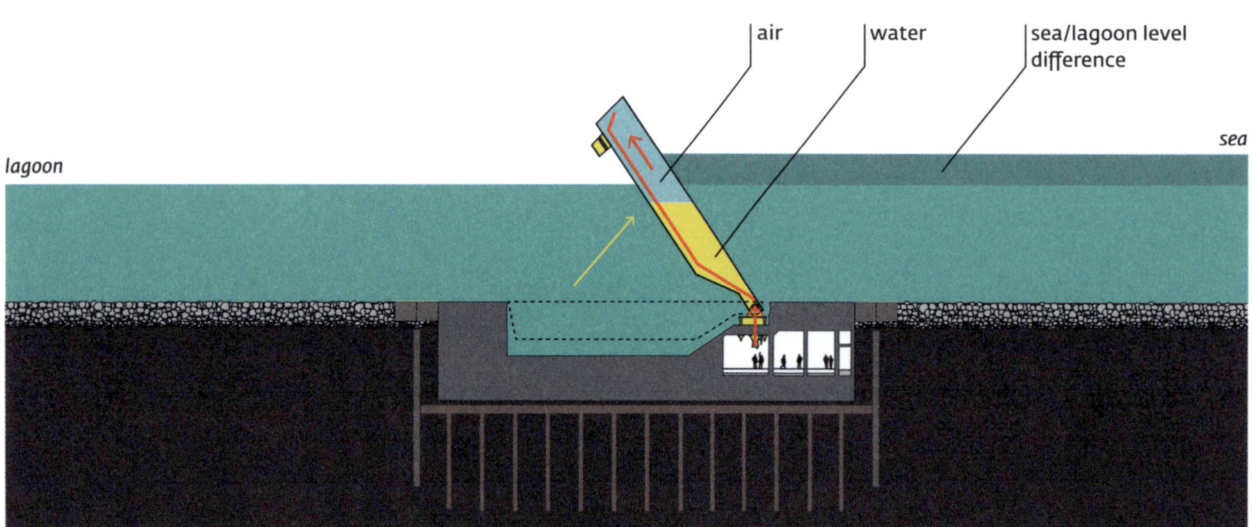

3. The lightened floodgates emerge from the water due to the buoyancy of “Archimedes principle” and rise, rotating around the hinge axis, to form a barrier

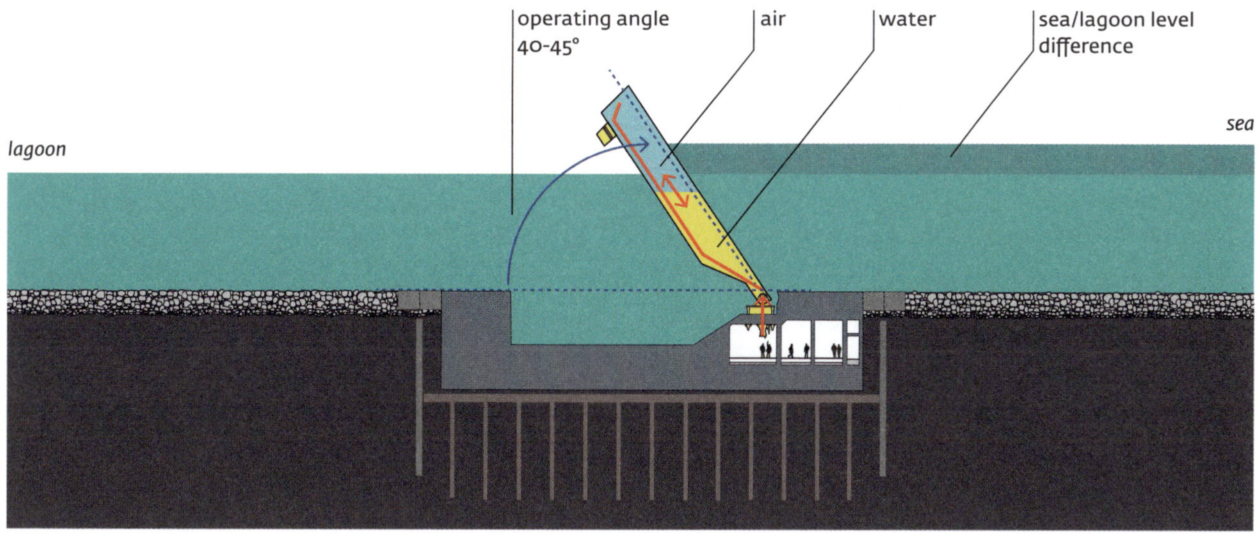

4. The floodgates are brought to and maintained at the optimum operating angle for the duration of the barrier lift, maintaining the difference in level between the sea and the lagoon. Subsequently, the trim of the floodgates is adjusted to adapt to changing tidal levels (so-called tidal tracking), again by introducing or expelling air

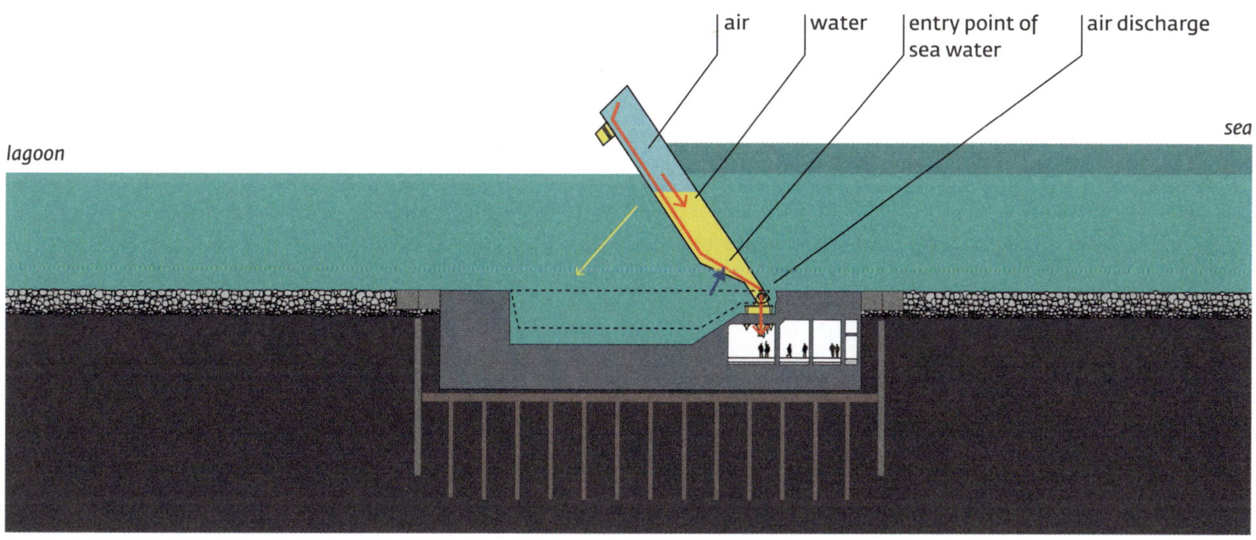

5. When the tide drops and the level between the sea and the lagoon is the same, the lowering phase begins. By discharging air, the floodgates are again filled with water and the gradual immersion phase begins

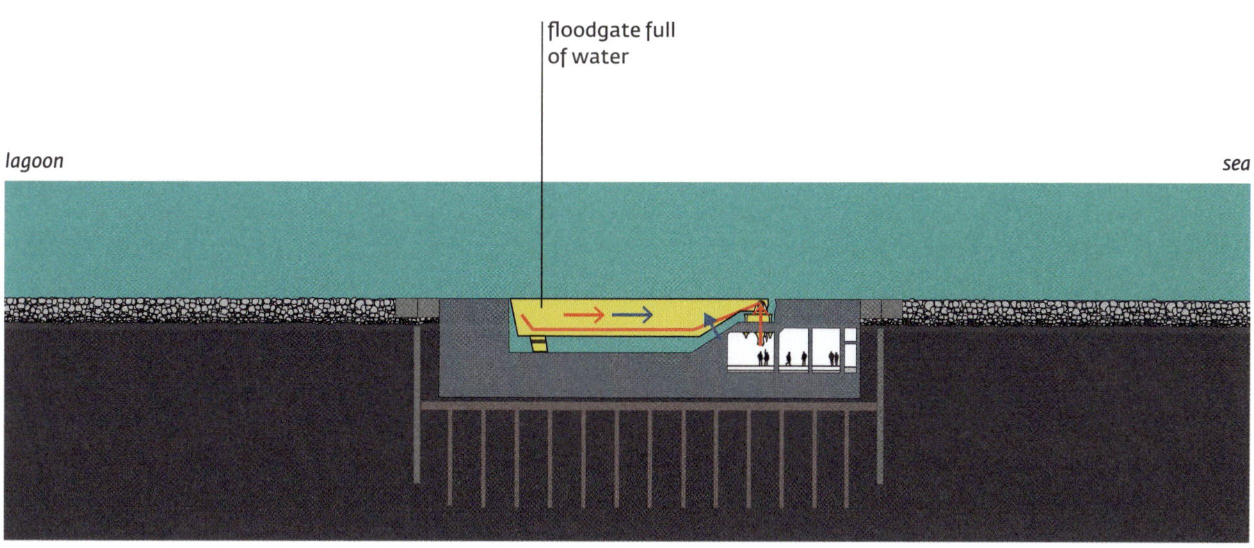

6. The water-filled floodgates return to their housing at the bottom of the entrance channels

Raising operations of the floodgates undertaken to date

The raising of the floodgates carried out to date can be divided mainly into functional tests in "calm" conditions and lifts in the provisional phase for the protection of Venice.
The former consist of all the tests carried out and that will take place at the barriers in "calm" weather and sea conditions, performed on individual gates and/or in groups. These tests are strictly necessary for the commissioning procedures of the plants, instruments, and control system and are also functional for the maintenance of the floodgates, for which the original project envisaged a timeframe of 30 days. The second category consists instead of all the raising procedures that have been carried out during the provisional phase to safeguard Venice, starting from autumn 2020. The provisional phase is the one preceding the completion of MOSE (scheduled for December 31, 2021).
Flood protection in this period has been and will be implemented in accordance with the guidelines issued by the Commissario Straordinario (Extraordinary commissioner) and the Provveditorato (Superintendent's office), which have already been shared with the other institutional bodies within the Coordination Committee, in the event of exceptional high waters and for all tides above +130 cm above the Punta della Salute zero mark.
Even though the works have not yet been completed, numerous tests have been carried out that demonstrate the functionality of this highly innovative system, thanks to which significant results have been achieved.

The means used for floodgate maintenance

The original maintenance project envisaged that the floodgates would be replaced using two special jack-up vessels working in parallel (one removing the gate to be maintained and transporting it to the construction site, and the other transporting the new gate and installing it in the barrier). It was envisaged that, if necessary, the replacement could be carried out with a single jack-up for the two operations, although this would obviously lengthen the time required. These vessels were also to have been available for the first installation of the gates.
Due to the delay in commissioning the first jack-up, the installation of the floodgates at Treporti was carried out with a different, specially designed construction vehicle: the gantry. This was cheaper and simpler, but slower, and required the use of other vessels and supporting infrastructure.
After the appropriate modifications for the different sizes of the gates and depths of the barriers, the gantry was used as a construction vehicle for the laying of the gates at Malamocco and Chioggia. Once it was found to be working well, it was decided to transform it from a construction tool to a maintenance one and to reinforce and test it for the same operating conditions as the jack-up.
With the adoption of two different means, the transfer operations take place in different ways to those envisaged in the original project: each floodgates is transported to the barrier by pontoon from the maintenance yard and vice versa; the transfer of each floodgates to the jack-up takes place at the lagoon entrance with a specific mooring; the transfer of the floodgates to the gantry takes place at the barrier, after it has been positioned over the barrier caissons and the pontoon moored.
The removal and repositioning of the floodgates required the construction of

↑ The gantry during installation of a flood gate.
↓ Two special means are used to install and maintain the gates: the jack-up, constituting the main tool, and the launching gantry as back-up. Installation of a floodgate for the Lido San Nicolò barrier.

ancillary works: mooring posts to moor the pontoon to the barrier when it is at the service of the gantry; moorings near the barriers to allow the transfer of the floodgate between the pontoon and the jack-up and vice versa (all in safe conditions).

The Mose I jack-up

The jack-up is a self-lifting C-shaped platform capable of lifting or installing the floodgate in its housing and transporting it floating to and from the maintenance site, without being affected by current and wave conditions.
The hull consists of two main elements, each L-shaped, with tapered ends, and an intermediate parallelepiped-shaped section, which allows two configurations: a short one for the Lido Treporti barrier and a long one for the Lido San Nicolò, Malamocco, and Chioggia barriers. The three components have an internal structure like a ship, with watertight bulkheads interspersed with longitudinal and transversal reinforced frames that divide the volume into a number of compartments, some used for ballast tanks, fresh water, and fuel and others for engine rooms or service areas.
The hulls are joined by means of a centering and coupling system located at the four tops of the contact surfaces, consisting of anchor plates blocked by wedges driven by hydraulic jacks.
The vessel is equipped with four square-section retractable legs that can position themselves and rest on the foundation caissons within special receptacles and support the weight of the hull and the sluice gate during the various installation phases. The legs are equipped with an articulated foot mounted on elastic supports for initial engagement and centering in the receptacles.

↑ → Details of the jack-up: in the dry dock at Arsenale Nord where it was assembled and during the installation of a floodgate.
↓ Steps in laying a floodgate using the jack-up vessel.

MOSE 1

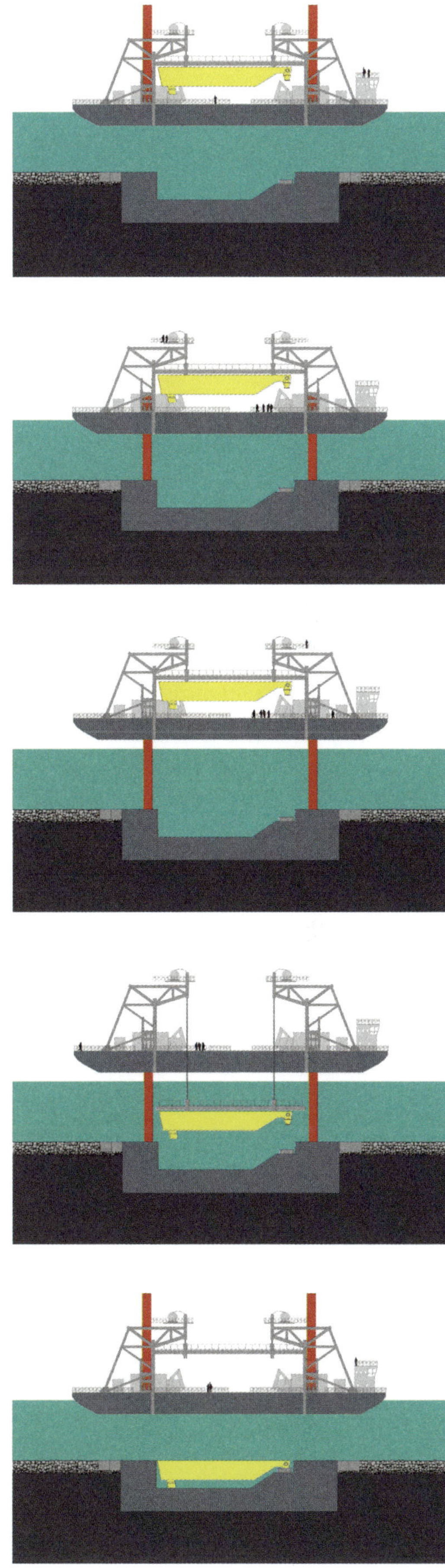

The floodgate is moved by a lifting system consisting of two "gantries" on which four strand-jacks are mounted. The floodgates are attached by means of a reticular interface structure, the "fisherman's frame," which also guides the floodgates into position. Special rotating connectors, mounted on the lower part of the frame, have the task of hooking the gate at four predefined points and supporting its weight during the recovery and repositioning phase. The fisherman's frame is able to connect the jack-up to the floodgate and allow precise planimetric adjustments to ensure the correct positioning of the sluice gate in its recess and the insertion of the male connectors into their respective female seats. Engagement is by means of twist-locks located on the fisherman's frame: these are housed in predetermined positions in the frame and at their locations, so that raising or lowering occurs on the vertical of each of the engagement points.

The engagement sequence involves lowering the twist-locks to clutch the floodgate attachment points and rotating them so that they engage in the slots of the attachment points, and finally raising them. Lifting can then be carried out. The release of the gates is done in a reverse sequence.

The vessel is equipped with a positioning system consisting of four water-jet thrusters, driven by electric motors and controlled during navigation and installation by a dynamic positioning (DP) system. It is also equipped with all the on-board systems necessary for navigation, vehicle operation and trim control and includes crew accommodation, as well as all the auxiliary systems for operational management and safety systems.

↑ A detail of the launching gantry: the "fisherman's frame," in yellow, connects the floodgate to the gantry and enables it to be removed and/or laid.
↓ Installation of a floodgate in the Chioggia barrier
• Installation phases of the launching gantry.

The jack-up is a very sophisticated vessel designed to operate with much tighter tolerances than those normally found in similar marine equipment (millimeters rather than centimeters). These tolerances were achieved by applying precision workshop technology to a marine vessel, a highly innovative solution that enabled the crew, after months of training and hands-on experience, to operate successfully.

The CV.EVO.02 straddle gantry and TP.EVO.02 transport girder

Because of the longer than expected completion and commissioning times for the jack-up, the laying of the Malamocco and Chioggia gates took place with the construction site system used by the company for the first laying of the Treporti gates, suitably adapted to the new dimensions. The worksite plant, as mentioned, was then acquired and adapted in order to increase its useful life, with the aim of using it for maintenance operations and making it to all intents and purposes interchangeable with the jack-up, to operate in the same conditions as the latter.
The straddle gantry too can operate in two different configurations: a "long" configuration for the Chioggia, Malamocco, and San Nicolo lagoon entrances and a "short" configuration for the Treporti entrance. The straddle gantry has a steelwork portal structure consisting mainly of two pairs of vertical lattice towers, two tubular guide towers arranged between the two lattice towers and two horizontal top girders supporting the upper platform.
Each pair of lattice towers is rigidly connected to the guide tower by two frames of girders called the base frame and the top frame respectively. At the mid-height on the towers there is a middle frame, which allows the articulation of both vertical elements (guide towers) to obtain the transport configuration with the lower part of the mobile legs rotated upwards.
The base frame is the interface element of the launching gantry bearing the floodgate caisson; it is equipped with a hydraulic damper system with cylinder and pressure accumulators, which accompany the laying of the gantry on the sea bottom and allow the self-centering of the same within the receptacles. The top frame consists of two box girders connected by transverse panels and acts as a connection between the CHB top girders with lattice legs and the guide tubes for centering the system.
The floodgate is moved by a lifting system consisting of four strand-jacks with cable recovery, mounted on four pairs of secondary girders which are in turn supported on the main CHB top girders. The attachment of the floodgate is affected through the interposition of a reticular interface structure, namely the fisherman's frame.
The centering function is assured by the shifting and rotation of the frame using the fisherman's frame; this mobile interface device is located on the guide towers of the launching gantry during the raising or lowering of the floodgate. On the straddle gantry there are also a lifting system that allows the rotation of the mobile legs for its movement on land and in the lagoon, a system for centering the gantry in relation to the foundation caisson and for moving the floodgate. The top platform contains the technical rooms: the control stations for the hydraulic system that lifts the legs and the floodgate, and handles the fisherman's frame, a workshop and engine room for the hydraulic control units serving the strand-jacks system,

PAGIOLI
ANGELINA C

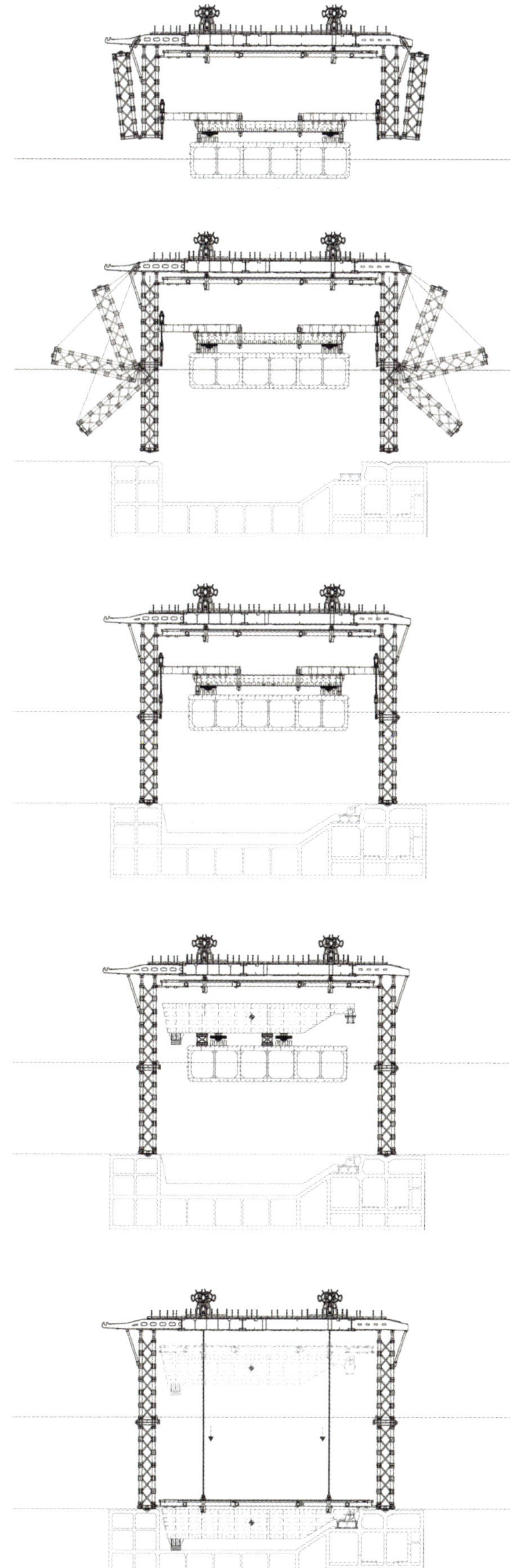

the electrical panel room, a changing room, and services. There is also a transport girder to be installed on the pontoon that allows the handling and transport of the gantry to the various barriers.
The TP.EVO.02 transport girder is a system of six telescopic girders equipped with four strand-jacks capable of raising and lowering the entire launching gantry; the two lower girders are welded box profiles 22 m long and constitute the fixed part of the structure.
By means of a dedicated hydraulic system, the four upper girders are able to slide in pairs on the lower girders and position themselves in three different configurations, depending on the operational need: a closed configuration, when the upper girders are completely retracted; a short open configuration, when the opening of the upper girders has a center-to-center distance of 25 m; a long open configuration, when the opening of the upper girders has a center-to-center distance of 34 m.

OLIO IDRAULICO

Barrier installations

The barriers are raised by feeding air produced by a system of compressors into the gates at a higher pressure than that in the floodgates at rest. The compressors and the entire air transfer system are 100% redundant and are located in completely segregated rooms and paths. Redundancy is the characteristic element of the entire system given the importance of the objective, the frequent use of the barriers (compared to other existing barriers) and the number of components (78 floodgates, 156 connectors, 312 units for transferring air to the gates, another 312 units for discharging air, valves and associated control and command instruments).
The systems required for operation are the automation and control system for the systems and gates at each lagoon entrance, the power supply to the utilities, the electrical building, the generator building, the compressed air production system, the compressor cooling system (air coolers), the air system for the instruments, the ventilation and conditioning system, the fire detection and extinguishing system, the anti-intrusion system and the telecommunications system.

Automation and control system for the floodgates at each lagoon entrance. The system for operating the gates is made up of numerous elementary systems, each of which has many components. This applies to both the main and redundant systems, which are identical and operate in the same way. The operation of such a complex system cannot be managed manually to cope with events that can occur even frequently and do so at short notice. A fully automatic system keeps all systems under control, even when they are at rest, in order to be able to put the flood defense into operation in the necessary time. In emergencies, the system can also be operated in semi-automatic mode, with the possibility of manually modifying the operational parameters. The operator can monitor all operations through a series of monitors dedicated to the individual functions from a control room located in the main abutment building at each lagoon entrance. In parallel and automatic mode there is also an ESD (Emergency Shut Down) procedure that secures the plant during barrier lifts.
The control system remotely manages the operational phases and the main functions of the lagoon mouth systems: Process Control System (PCS), Emergency Shut Down (ESD), and Power Management System (PMS). The system as a whole consists of three sub-systems installed at the Lido, Malamocco, and Chioggia sites, each of which transmits information on its status and receives operating commands from the relevant lagoon entrance PCS. The three entrances are interconnected by a dedicated fiber optic ring and radio link.
In terms of hierarchy, the superior station is the centralized one located in the automation and control building at Lido San Nicolò, which gives instructions for raising and lowering the gates to all the other control rooms. There is also a further remote centralized station at the Arsenale (Tesa 112), which is a replica of what is planned for the Lido inlet. Thus articulated, the system allows the movement of the floodgates, the management of compressed air production and the drainage system, and the flushing process. It also monitors the HVAC and fire-fighting auxiliary systems, the connector-hinge coupling/

→ The area with the technical buildings on the Lido.
↓ Ducts for the passage of compressed air to the floodgates in the underwater tunnels.

ARIA DI PROCESSO

uncoupling system, the watertight doors of the tunnels and the locks.
There are also two “controllers” per caisson (A and B) of the same process elements to allow for the redundancy of the plant and field systems.
The ESD process control system intervenes by pressing the emergency button by a suitably trained operator following a situation of instability of one of the gates, reducing the working angle of the barrier, or because of an excessive air bubble pressure inside a gate. The ESD system also requires two redundant “controllers” (A and B) per caisson.

Electrical power supply to utilities. The maximum electrical power absorbed by the plant during the transitional phases is 13.5 MW (Lido 6 MW, Malamocco 4 MW, and Chioggia 3.5 MW). Each system is developed using the following voltage levels: medium voltage 20 kV for connection to the ENEL network; medium voltage 6 kV (as an alternative to the ENEL power supply, 6 kV emergency power can be produced using 2+2 diesel generators); low voltage 400/230 Vac for process auxiliaries and general services; low voltage 230 Vac for powering the control system and all auxiliaries.

Electrical building. The electrical building planned for each lagoon inlet houses the transformers and switchboards for the 20kV/6kV and 6kV/0.4kV exchange. For each section there are: a resin transformer, UPS, medium voltage switchboards; low voltage main switchboard; cooling and ventilation units for the electrical rooms; hoists, stairs, fire-fighting systems, etc.

Generator building. The emergency power generators, the switchboards for their operation, the accessories and the air conditioning system are all housed in the generator building located at the main abutment of each barrier. The building is divided into two wholly separate rooms, in each of which the emergency generators are installed. The power output available is 2000 kW at Malamocco and Chioggia, 2500 kW at Lido. Also installed here are the daily diesel storage tanks for each generator and those for the gravity collection of the diesel, electrical panels, equipment for the ventilation of the rooms and the cooling of the control rooms, and fire-fighting systems.

Compressed air production plant. The compressed air plant comprises 3+3 centrifugal compressors, located in two wholly separate rooms. The discharge pressure of each compressor is 4 bar absolute. The flow rates are 8000 Nm^3/h at Lido, 6000 Nm^3/h at Malamocco, and 5000 Nm^3/h at Chioggia. In the same building there are the electrical command and control panels for the compressors; the auxiliary systems for the wall outlets, the ventilation of the compressors, the lighting and their related panels; the cooling and ventilation units for the rooms in the building; the fire prevention systems; and the goods lifts.
The plant also includes the cooling system and compressed air delivery piping. The pipelines are located in underground tunnels up to the abutment. From here, through shafts in the building, the pipelines reach the two tunnels and, in progressive sequence, the connector rooms and the floodgates.

Compressor cooling system (air coolers). The compressor cooling plant is located at

each main lagoon entrance abutment. Each section comprises 6 air coolers, the brine circulation pumps, their control panels and systems, and the circuit expansion vessels.
The two redundant sections work as a back-up to each other but can operate independently to cool any compressor if required.

Air system for plant. The plant equipment includes the system for the production and distribution of compressed air for the distribution of drinking water and services, for the operation of emergency breathing apparatus in the tunnels, for cleaning and other needs in the rooms, for the instruments, for the operation of local compressor panels and pneumatic valves. The service air is only pressurized, while the instrument air is also dehumidified.

Ventilation and air conditioning system. All the rooms where the staff work and where the systems are located are equipped with ventilation and air conditioning (HVAC) to keep the humidity level low and limit changes in temperature and are sized according to the specific needs; in each of the buildings there are rooms for the HVAC machines, their electrical connections and the control panels. The distribution of cooled air is affected as in a normal civil building, but ensuring the segregation of redundant rooms.
The tunnels and technical rooms have their own HVAC system, with built-in redundancy and installed on the roof of the abutment buildings: half the tunnel is cooled by one abutment, the other half by the opposite abutment.

Fire detection and extinguishing system. All the rooms where the systems are located

↑ → Some of the floodgate installation phases and assembly of the air pipes from the tunnels through the male element of the hinge to the floodgates.

and where staff work, even if only occasionally, are equipped with sensors. The fire extinguishing systems, which have built-in redundancy, are different according to what they are designed to defend, and include: a network of hydrants for external protection of the buildings, foam systems for rooms or areas where combustible liquids are present, water mist extinguishing systems for the tunnels in the barrier caissons and underground tunnels, inert gas extinguishing systems for switchboard rooms, instrument rooms, control rooms, emergency rooms and transformer rooms,
and fire extinguishers.

Intrusion and telecommunications systems. The systems also include anti-intrusion, video surveillance, and access control systems and a telecommunications system consisting of internal and external lines, local intercom with loudspeaker call, Ethernet network for computer interconnection, and a radio system for each lagoon entrance to ensure communications from the tunnels in any situation.

The MOSE System Operations Center

The MOSE System Operations Center is the set of functions undertaken by the operations manager and the plant manager within the decision-making operations room, with an annexed emergency management room and the remote centralized control room.
In technological terms, the automation and control system is part of the latter: located at each lagoon entrance, this is the system that communicates on a computerized level with all MOSE components to receive diagnostics and issue commands and that ensures the operation of the barriers. The system consists of a Process Control System for each lagoon entrance, which receives and sends information to and from each component of the system connected to it, and a room where the personnel in charge monitor the progress of operations and intervene when necessary. The personnel are divided between those who supervise the operations and those who carry out continuous inspections and routine maintenance.
The three automation and control systems are operationally dependent on the Central Operations Center, which is located on Isola Novissima, at the mouth of Lido.
Through the plant manager, the Operations Center effects the maneuvers ordered by the decision-making Operations Room, manages emergencies through the Situation Room, and carries out maintenance activities on the plants, IT equipment, and control rooms.
As a whole, therefore, the Operations Center controls the entire flood defense system, monitoring the status of the plant and the meteorological parameters. On the basis of these, it takes decisions regarding the operation of the flood defense gates and also as concerns ordinary and extraordinary maintenance, maintaining contact with the stakeholders, bodies and managers of activities that are affected by the operation of the barriers. It is also responsible for monitoring the lagoon environment that may be affected by the operation of the barriers.
The Operations Center is located at the Arsenale and manages the flood defense in complete autonomy by operating the barriers on the basis of operating rules; that is, in line with precise protocols established at the time by the Decision

← ↓ The northern area of the Arsenale.

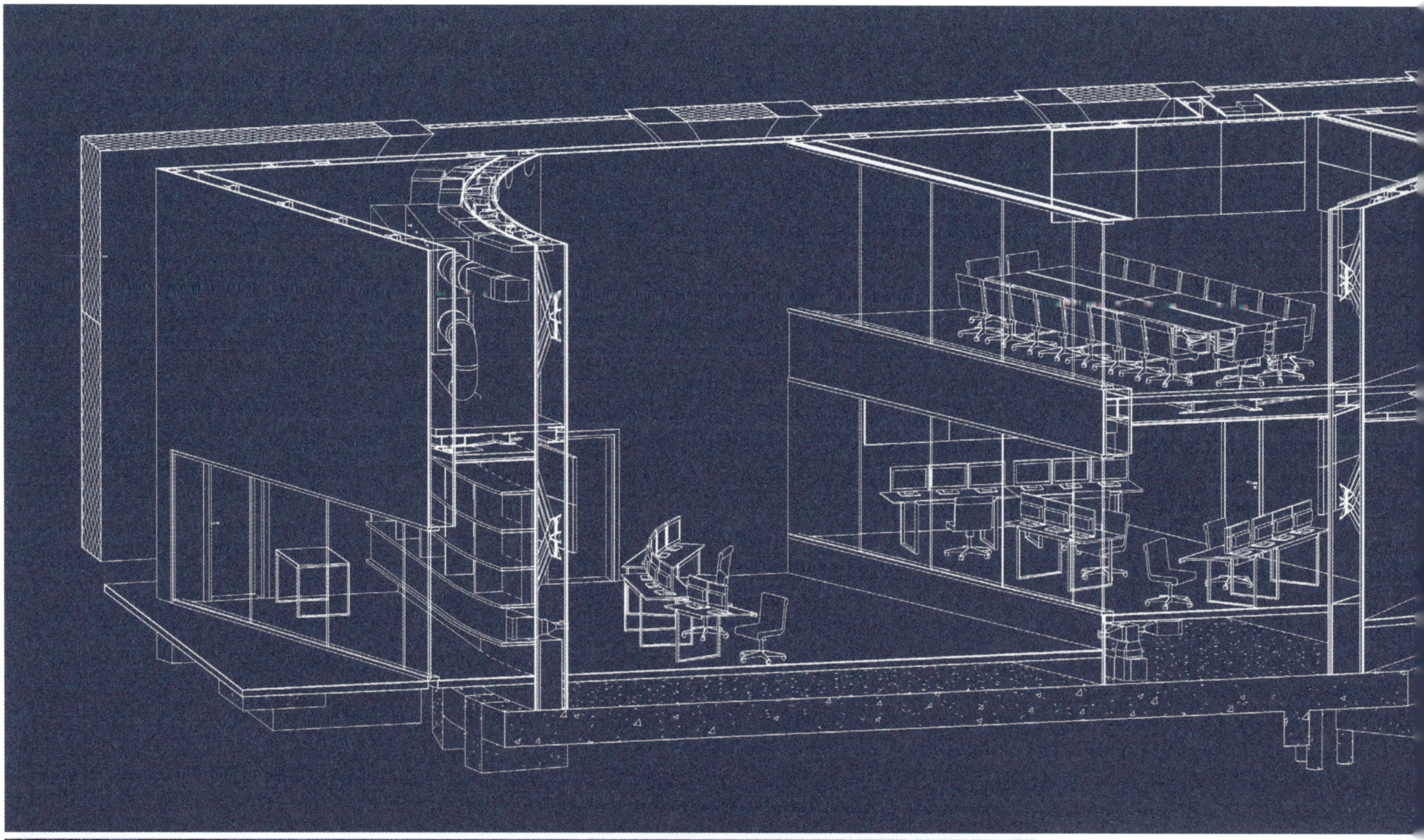

The General operations center for the activation of the barriers at Arsenale Nord
↑ The Decision-making operations room
← The backup room

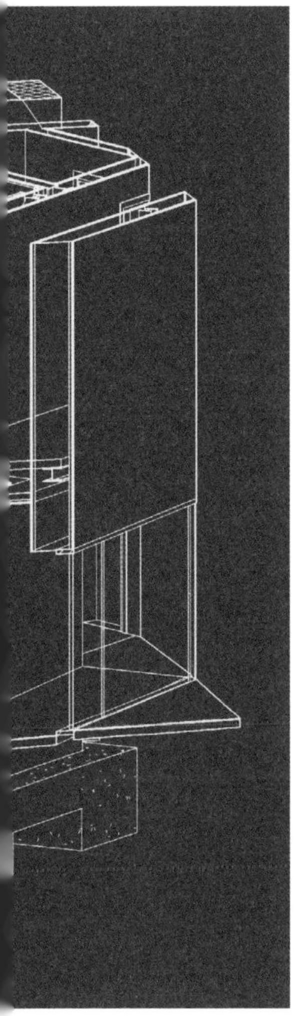

Center comprising the Provveditorato and the Commissario straordinario per il MOSE, after consulting the local authorities with updates following the results of the unplanned and unprogrammable lifting tests during high waters in the provisional phase.
The procedures will need to be updated over time on the basis of experience, with periodicity borrowed from other similar situations of management of large flood defense barriers.
The operating procedures for the provisional and definitive phases are applied directly by the Operations Center, since the decision regarding the movement of the floodgates must be taken in a short time (in a matter of hours, given that at the moment the works are still in a provisional and construction phase and the works have not yet been completed and tested).
When fully operational, the Operations Center also interfaces with the "Crisis Unit" in emergency conditions. The Operations Center therefore has lagoon entrance control rooms and related Data processing centers, the Operations control room, Situation room, Decision-making operations room, and related Data processing centers (in the Arsenale), rooms for the operation and management of the decision-making support system, where the measurements of the monitoring networks converge, information from the forecast systems (in the Arsenale), and management systems for maintenance, for the purpose also of periodic updates of the operating rules.
At a technological level, the infrastructure serving the plant management service consists of Server Room A and Server Room B at each lagoon entrance, as well as a Lagoon entrance control room. The Lido Lagoon entrance control room also acts as a centralized station control room. Signals from the automation and control systems are sent to it, and from the control rooms the commands for raising and lowering individual gates and entire barriers can be sent.
At present, the connection between the barriers is provided by a radio link supplied by the Italian Army and when fully operational will be assured by a double fiber optic ring connection as primary communication channel and by a backup communication c hannel using the above-mentioned radio link.
In the Arsenale, the infrastructure serving the decision-making operations room of the MOSE System consists of two production data processing centers, operating in unison through MetroCluster storage with synchronous updating.
Each data processing center (CED, to use the Italian acronym) in the Arsenale is equipped with a double power supply line with redundant internal electrical distribution systems. An additional CED is available, which becomes operational in case of collapse of the first two, following a specific activation procedure called "IT Disaster Recovery Plan."
The power supply is guaranteed upstream by two UPSs, as well as by two generators. Similarly, the air-cooled chilled water conditioning systems are redundant, each with a capacity of 205 thermal kW, also equipped with free cooling batteries.
The CEDs contain rack cabinets equipped with wired switches per cabinet, high-performance storage systems, and physical servers for a total of 448 physical processors per hundred, in

order to virtualize all the servers required for operations. The entire system is protected by specific security equipment. The decision-making operations room has five workstations with three screens, through which operators interact with the virtualized systems and with the outside world (sending messages and maneuver orders). In the room there are two panels with six screens each, which ensure the representation of aggregated information, which is useful to operators in the decision-making phases.

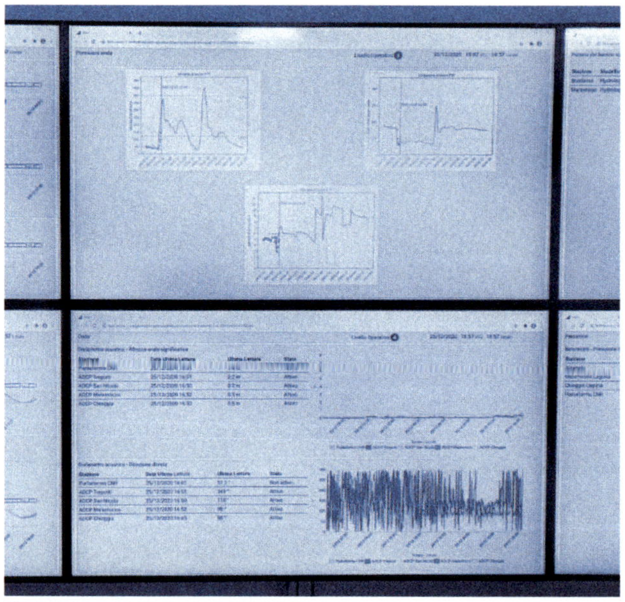

Lagoon entrance control rooms and associated data processing centers

A control room is foreseen at each barrier alongside the lagoon entrances. A Server Room A and a Server Room B will be dedicated to this as local data processing centers.
The control room of each lagoon entrance receives continuous information on the status of the individual systems of the entire barrier: status of the water, glycol and industrial systems flows; status of the power supply, fire prevention systems, intrusion detection, video and telephone systems, plant air system, drainage and condensation, HVAC, generators, watertight doors, telecommunications and navigation locks.
The control rooms of each lagoon entrance operate the local barrier, lifting and lowering the floodgates based on the indications given by the decision-making operations room, and collecting all the information relating to the behavior of the individual plant components. At the control room of each barrier, assessments are carried out regarding the degree of functionality of the system, any temporary fault in some redundancy or nonfunctioning position of a floodgate, and indications regarding the replacement of components and supplies.

↑ → Details of the control and verification systems for lifting operations and the operations room on the island at the Lido entrance. Each inlet is an entrance with its own independent and locally managed control system with a manual control console for moving the floodgates. The control rooms at the entrances operate the movement of the barriers on the basis of indications issued by the Decision-making operations room at the Arsenale

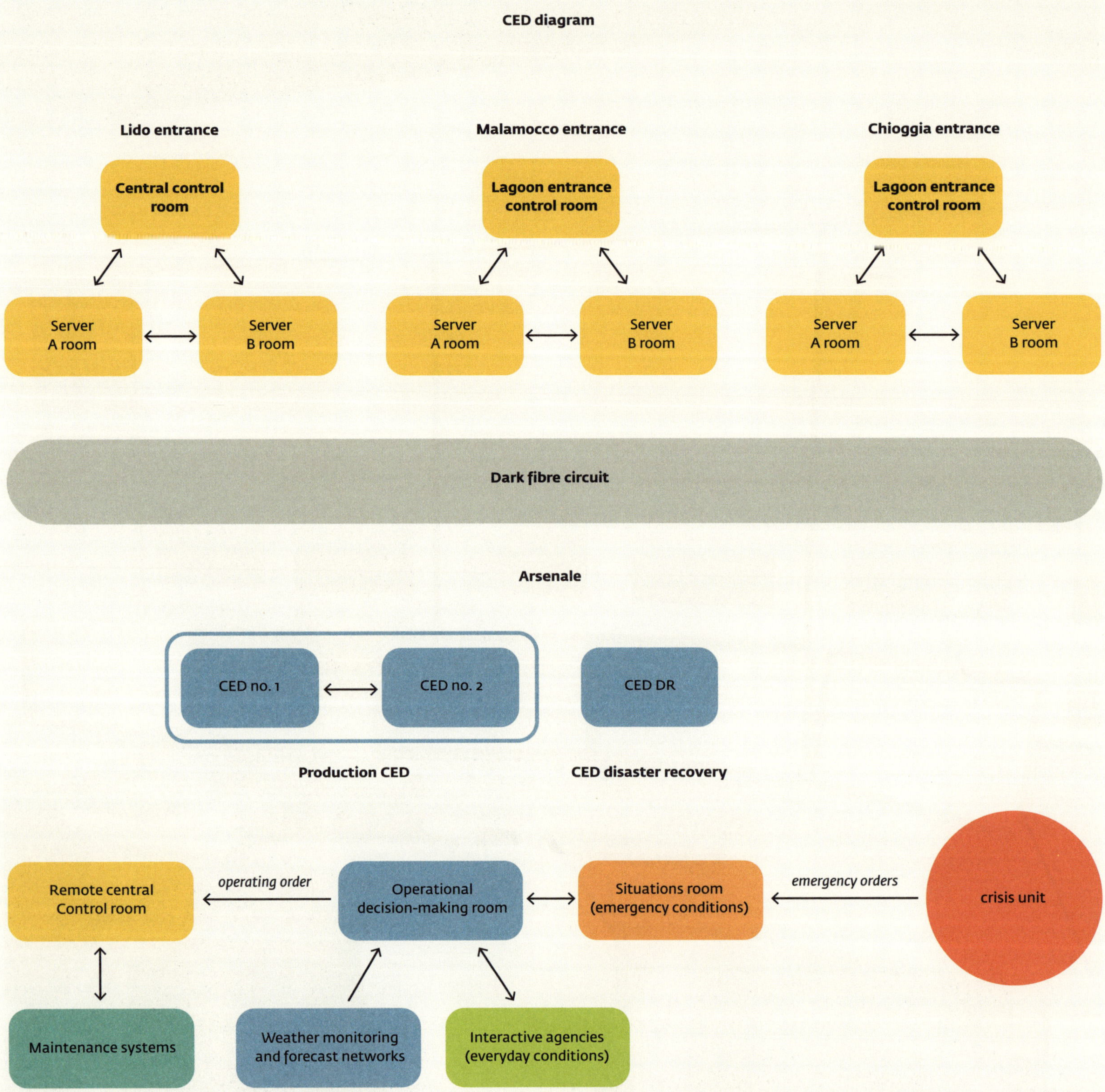

Decision support system. Above, information flowchart; below, classification of highwater events.

The results of these evaluations are transmitted to the Lido control room and, from there, to the control room in the Arsenale via a single-mode optical fiber that has already been laid. The aggregate status of the systems will be made available through the automation and control system currently being developed as part of Tender 32 and possible additions. The control rooms are still being completed and are being operated temporarily from temporary rooms. The same applies to server rooms A and B at the lagoon entrances.

Operations control room, Situation room, Decision operations room, and related Data processing centers

Once completed, the Operations center at the Arsenale will have an Operations control room, a Decision operations room, a Situation room, and related Data processing centers. To date, the backup Decision operations Room is available and has been built and used for the management of high water events in autumn–winter 2020–2021.

Monitoring networks, forecasting and decision-making systems

In order to issue orders to move the floodgates (lifting and lowering) and to define the operational weather and sea conditions windows for MOSE maintenance activities, the Operations center uses a series of monitoring, forecasting, and decision support tools. In view of the very limited time available to make arrangements, the center must operate in complete autonomy on the basis of operating rules agreed in advance and periodically updated through a control room.
The Operations center has meteorological and marine monitoring networks at sea and in the lagoon managed by a monitoring network unit: data on levels, tides, wind, and rainfall are required to feed the decision support system, while data on wave motion and currents are needed to organize maintenance of the floodgates. All this information, together with information on the quality of the lagoon environment and the activities taking place there, is indispensable for evaluating the effects produced by operation of the barriers.
The measuring instruments consist of detection sensors powered by solar panels and, where possible, a wind generator, when power from the national grid is not available. The monitoring stations are equipped with buffer batteries and multiple data transmission systems and, at critical points, with two different sensors that assure the operation of the weather and sea sensor and provision of data (in particular with regard to sea levels) using two different technologies. Over the years, data exchange agreements have been set up with agencies and weather centers (ARPA SIMC, Reading, NOAA, ARPAV), some of them free of charge.
The decision support system uses measurements and forecasts of sea level and duration above the tide threshold in the sea and in the lagoon, of precipitation in the lagoon and on the drainage basin with relative flow, of wind in the sea and in the lagoon to define the overflows generated.
Technologically, the system is highly configurable and allows operators to modify the rules implemented directly from a Windows form; it updates at 5-minute intervals on the basis of data and forecasts from Punta della Salute and Chioggia Vigo and defines at what level the closure of the inlet channels must be initiated to ensure protection of the lagoon settlements against flooding.
Sea level forecasts for the upper Adriatic Sea are generated with autoregressive statistical models (on predicted sea levels

and pressure gradients), propagated in the lagoon with a two-dimensional hydrodynamic model to provide forecasts in Venice and in the historic center of Chioggia. In the future, statistical autoregressive models may be replaced by deterministic models currently being developed and tested, which operate with real-time data assimilation techniques in the Adriatic. In addition, a system for forecasting the flow of water from the drainage basin is already available with an inflow-deflow model based on rainfall in the drainage basin. The forecast error is also taken into account in the operation and is automatically used in the decision support system to define the instances of lifting and lowering of the barriers with an appropriate safety margin.
Additional indications for operators regarding the state of health of the environment come from satellite images, information from the lagoon and the sea such as the warning bulletins, from synoptic forecasts and weather warnings from the Dipartimento della Protezione Civile (Civil protection department), from the decentralized functional center of the Veneto Region, and from the Friuli Venezia Giulia Dipartimento della Protezione Civile, etc., which are displayed on the video walls of the decision-making operations room and the situation room.
Information on the aggregate status of the facilities will be transferred to the Decision-making operations room from the lagoon entrances via the Operations control room: this information covers the status of the main systems to summarize the aggregate operational status; anomalies on the positioning of some gates and on the operation of the lagoon entrance plant.
From the Decision-making operations room to the Works control room, information will be transferred concerning: the operational level at all times of the day, every day, on five levels with the relative color scale; the values of the levels reached above which the barriers will be closed.

→ The backup room

Tide forecasts and management

Stefano Libardo
Alvise Papa
Davide Sernaglia

The tide within Venice's Lagoon

← Venice, high water on December 1, 2008.

In over 1,600 years of history, the city of Venice and the lagoon settlements have always been subject to the phenomenon of high water. Chronicles show the first exceptional phenomena as early as 589 CE, but it is only since 1867 that the sea level has been scientifically and systematically measured. In Venice, tide levels still refer to the "Zero Mareografico" at Punta della Salute (ZMPS) defined by the State Altimetric Network of 1897, obtained as the average sum of the values recorded in the period 1885–1909. Since 1871, the official tide level in Venice has been measured using the same technique and in the same area. For the last forty years, the official value has been measured by instruments positioned in the tide gauge booth at Punta della Salute on the Grand Canal side.

In order to understand the complexity of the forecasting systems currently active in Venice, it is useful to present a brief view of the fundamental characteristics of the tidal phenomenon. The Adriatic Sea has a physical peculiarity that allows us to consider the tidal level as a linear sum of single components. This small facilitation provided by nature helps us to summarize in this text what the main factors are that characterize the sea level in Venice and its lagoon.

The main element is undoubtedly provided by the astronomical tide, that is the result of the gravitational forces that the celestial bodies—especially the Sun and the Moon—exert on the mass of water covering the Earth. Various methods can be used to calculate the astronomical tide, and the one used in Venice is based on the principles of harmonic analysis, which can describe the trend of this component for several years to come and with extremely small errors, generally of the order of a centimeter. The characteristic values of the astronomical tide in Venice see a maximum of +87 cm ZMPS and a maximum range of 116 cm.

A second component is described by the phenomenon called "seiches." The Adriatic Sea can be compared to an "almost closed" basin, and with this geometry special events such as impulsive depressions or strong winds that suddenly cease can give rise to resonance waves with well-defined periodicity but variable amplitude depending on the surrounding conditions: seiches.

Generally, the most important element in determining the tide level in Venice is given by what is defined here as the "meteorological contribution" to the tide, that is, the effect on the level of variations in atmospheric pressure and wind. Atmospheric pressure acts in various ways, the first in importance being the phenomenon of the "inverse barometer" or the relationship that exists between pressure and the compressibility of water: if pressure decreases, sea level locally increases. The theory tells us that for every hPa of variation in 24 hours, there is an opposite reaction of the level of about 1.5 cm on average. The second is associated with the shape of the Adriatic basin: if the atmospheric pressure is higher in Bari than in Venice, the water is forced towards the northern part of the Adriatic. Then there is the wind, which is certainly the most important forcing factor. In this case too, the morphology of the basin is fundamental: a long, narrow channel surrounded on three sides by high mountain ranges (Apennines, Alps, and Dinaric Alps) with a fetch of about 900–1000 km along its axis—oriented from south-east to north-west—very deep in the southern part with the South Adriatic fossa exceeding 1200 m and only 30/50 m in the northern part. With this morphology, intense winds, especially from the south-east—the "scirocco"—can cause a significant rise in sea level in front of the Venice lagoon. In recent history the highest meteorological contribution was recorded on November 4, 1966, with a value of +194 cm. Opposing phenomena such as high pressure, northerly winds, and so on can give opposite phenomena: in 1986 these led to a record negative meteorological contribution of -67 cm.
Within the Venice lagoon, the tide level

Classification of tidal events in the Venetian lagoon

Exceptional high tide
≥ +140

Extremely high tide
+110 ÷ +139

Extremely high tide
+90 ÷ +109

Normal tide
-50 ÷ +89

Tide below normal values
-90 ÷ -50

Exceptional low tide
<-90

→ The Lido San Nicolò barrier in operation during a high tide event, October 3, 2020
• The integrated defense system to protect the territory from high water consisting of local defenses and mobile barriers.

Venice
lagoon entrances
lagoon
local defenses
MOSE system

is the result of the propagation of the tide present in the northern basin of the Adriatic through the three lagoon entrances, which are now subject to the regulation of the MOSE system, and of similar phenomena occurring within the lagoon, although with somewhat reduced amplitudes.

Historical and recent *acqua alta* (high water) events

Since records began in 1872, there have been 25 episodes above +140 cm ZMPS; 55 above +130 cm ZMPS, and 326 greater than or equal to +110 cm ZMPS. The worst high water event for Venice occurred on November 4, 1966, with +194 cm ZMPS, followed by the recent one on November 12, 2019, which reached +187 cm ZMPS. In order to understand the urgency of providing the city and its lagoon with a protective system, it is also worth comparing the average number of tidal events between the periods 1966–2000 and 2001–2020. Due to the combined effect of subsidence and eustasy (see chapter 3), the average number of events greater than or equal to +110 cm ZMPS has increased from 2 to 4 per year. Moreover, the number of events greater than or equal to +130 cm ZMPS has increased from 1 every 2.5 years to 1 every 1.5 years; even more important is the value of tides greater than 140 cm, which have gone from 1 every 9 years to 1 every 2 years.

The situation regarding the lagoon settlements

The built-up area of Chioggia is protected up to +130 cm ZMPS by a protective system consisting of local elevations and mobile barriers on the Vena canal that prevent flooding above +82 cm ZMPS, while above +130 cm ZMPS protection is assured by the MOSE system at the lagoon inlets.
Venice and the islands of Murano and Burano are protected up to heights in

↑ → Venice, high water on November 12, 2019.

some cases lower than +110 cm ZMPS because a part of the local defenses has yet to be completed to reach that level. At present, with tides of +110 cm ZMPS even in the historic center of Venice, about 12% of the accessible public land is flooded.

Safeguard level

Institutionally, a safeguard level of +110 cm ZMPS has been established for the historic city center of Venice and +130 cm ZMPS for the historic center of Chioggia, as set out in the 1981 Feasibility Study—Outline Project (the so-called Progettone, or Megaproject, defined on the basis of the 1975–1978 tender) and in vote no. 209 of the Consiglio Superiore dei Lavori Pubblici (Higher council of public works) of 1982. The same level was maintained for the provisions of the Consiglio dei Ministri (Council of ministers) of March 15, 2001, for plans to raise the floors to maximum levels compatible with the historical, architectural, monumental, and landscape context (in any case keeping to a level of 120 cm), as reiterated at the Comitatone (Megacommittee) meeting of December 6, 2001, in the final design of the MOSE system of September 30, 2002, and at the Committee meeting of April 3, 2003. The safeguard level does not represent a physical limit to the system's operation: MOSE can also operate at lower heights, which increases the average number of annual closures.

The operational management of the MOSE System

The decision as to whether to raise the barriers is made on the basis of meteorological and marine measurements and weather forecasts, acquired by a software system that makes them available to the forward planning systems and from these to the decision support system that defines the closure time, the reopening time and the closure height foreseen for the event to be dealt with.

The decision-making support system classifies high water events in line with a closure strategy for full operation derived from the MOSE project and subsequent detailed studies and analyses.

The decision-making support system also generates notices to be communicated to the Port Authorities of Venice and Chioggia and other agencies, in line with the procedure for the interim phase starting 48 hours before barrier-raising. In its current version, this system is able to monitor and follow events as early as 120 hours before possible activation, depending on the availability of tidal forecasts.

Tide forecast for operational management

The forecast of the tide level is obviously the fundamental input for the activation of the MOSE system. Therefore, due to its importance, the concessionaire has been developing part of its modeling since 1995, at that time also in collaboration with the Centro Previsione e Segnalazione Maree (Tide prediction and reporting center). Since then, the development activities of the two organizations have proceeded in parallel, with the concessionaire collaborating with the Danish Hydraulics Institute and the Tidal Centre with CNR-ISMAR. An interesting Memorandum of Understanding has recently been undersigned that relaunches the collaboration between the concessionaire and the Tide Prediction and Reporting Centre, with the aim of sharing knowledge on tidal forecasts for the lagoon, but also with the aim of improving them and reducing the margin of error over time.

The Centro Previsione e Segnalazione Maree—active since 1962 and in its current form since 1981—has developed two networks of models: one statistical and one deterministic. The first calculates the tidal value at sea and independently in the lagoon; the second, deterministic, calculates the tidal value on an unstructured grid covering the entire Mediterranean area, onto which a modelling structure for the interior of the lagoon is then grafted.

The modeling structure operating at the concessionaire generates a forecast starting from the sea level in front of the lagoon entrances, using autoregressive statistical models, and is propagated in the lagoon using a two-dimensional deterministic model.

While waiting for the Memorandum of Understanding to bear the first fruits of collaboration and joint development, for operational management purposes an analysis of the data and forecasts available from the concessionaire over the last twenty years has been carried out, identifying forecasting errors by considering the 5th percentile in underestimation, comparing the forecast and reality in relation to individual tidal peaks. In the decision support system now in operation, errors in underestimation were therefore assumed, which only in 5% of cases can be worse, ranging from 35 cm at 48 hours after raising the barriers up to 10 cm at 1 hour after raising.

Today, at the time of raising the barriers, we proceed on the basis of the levels measured and not those predicted, thus further reducing the margin of error, despite the fact that communications to the Port Authorities initially take place only on the basis of the tide forecasts. Changing the degree of risk assumed (that is, the percentile value) changes the number of mobilizations of operational

personnel and increases or decreases the risk of not moving them when necessary or mobilizing them unnecessarily. Any decision changes can only be made after an adequate number of movements during operating conditions in order to fully assess the benefits and risks of the proposed change.
The interim operational activities carried out in the autumn–winter of 2020–2021 enabled an analysis of the accuracy of the forecasting systems available. This showed that the forecast system, and in particular the value of the meteorological component, could be improved. The analysis confirmed that the meteorological contribution is the most dynamic part of the forecast and therefore also the one that is most sensitive to slight variations in the surrounding conditions. Thus, it has been shown not only that the average error in forecasting sea levels in front of the Lido entrance is proportional to the amplitude of this component, but also that there is a relationship with its speed of variation. The slower the increase in the meteorological contribution, the lower the error; the more dynamic the phenomenon, the greater the error. On the basis of these data, it is possible to correct the prediction of tidal level at sea in front of the entrances and this possibility, already tested in similar solutions put online for the Dutch flood barriers, will be tested off line in the next high water events.
Improving forecasting and reducing the margin of error involves a complex process of analyzing past events and identifying improvements in the accuracy of weather forecasts, with a decades-long commitment from national and international meteorological bodies, and corrective measures applicable to

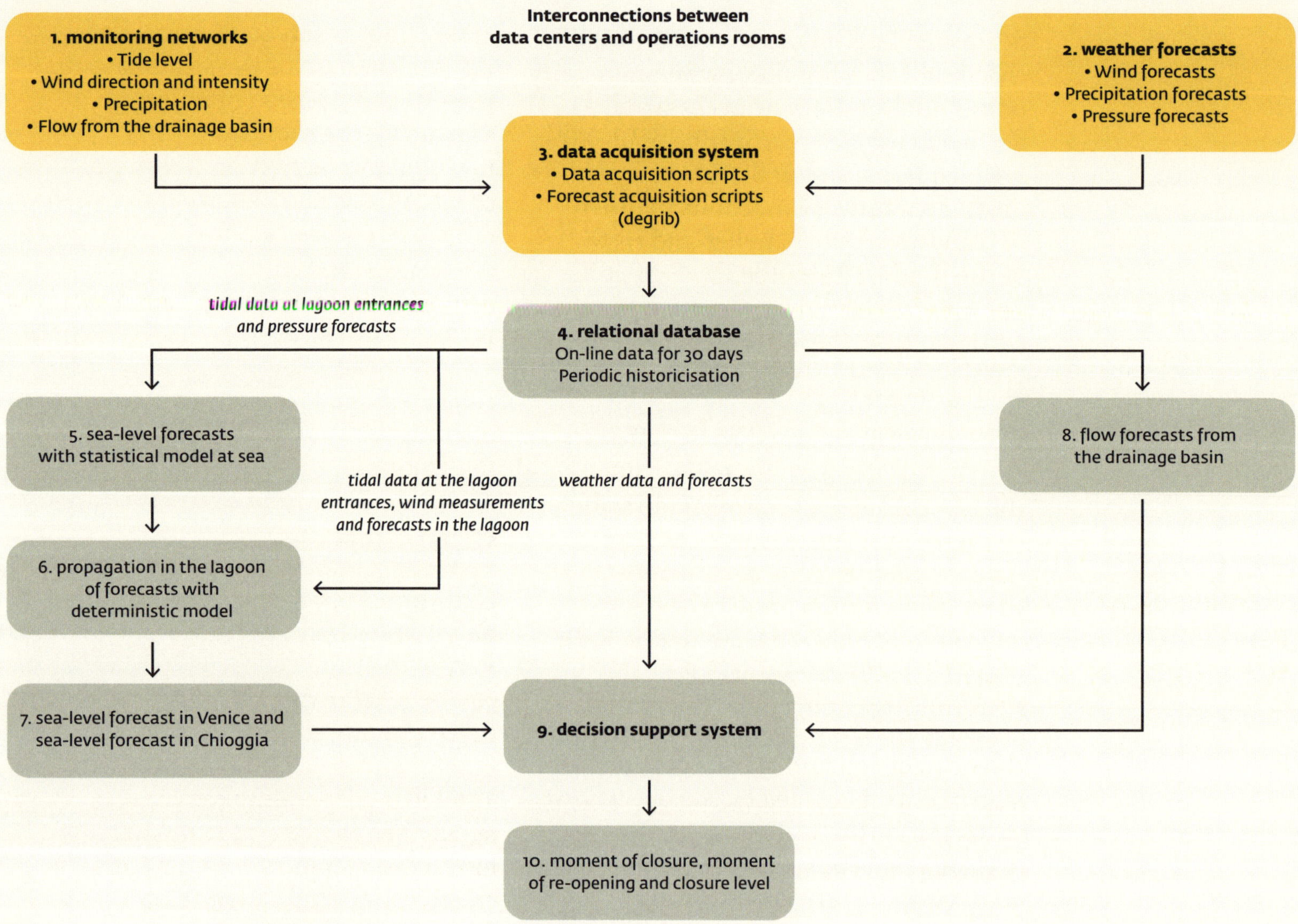
Interconnections between data centers and operations rooms
1. monitoring networks
• Tide level
• Wind direction and intensity
• Precipitation
• Flow from the drainage basin
2. weather forecasts
• Wind forecasts
• Precipitation forecasts
• Pressure forecasts
3. data acquisition system
• Data acquisition scripts
• Forecast acquisition scripts (degrib)
4. relational database
On-line data for 30 days
Periodic historicisation
tidal data at lagoon entrances and pressure forecasts
5. sea-level forecasts with statistical model at sea
tidal data at the lagoon entrances, wind measurements and forecasts in the lagoon
weather data and forecasts
8. flow forecasts from the drainage basin
6. propagation in the lagoon of forecasts with deterministic model
7. sea-level forecast in Venice and sea-level forecast in Chioggia
9. decision support system
10. moment of closure, moment of re-opening and closure level

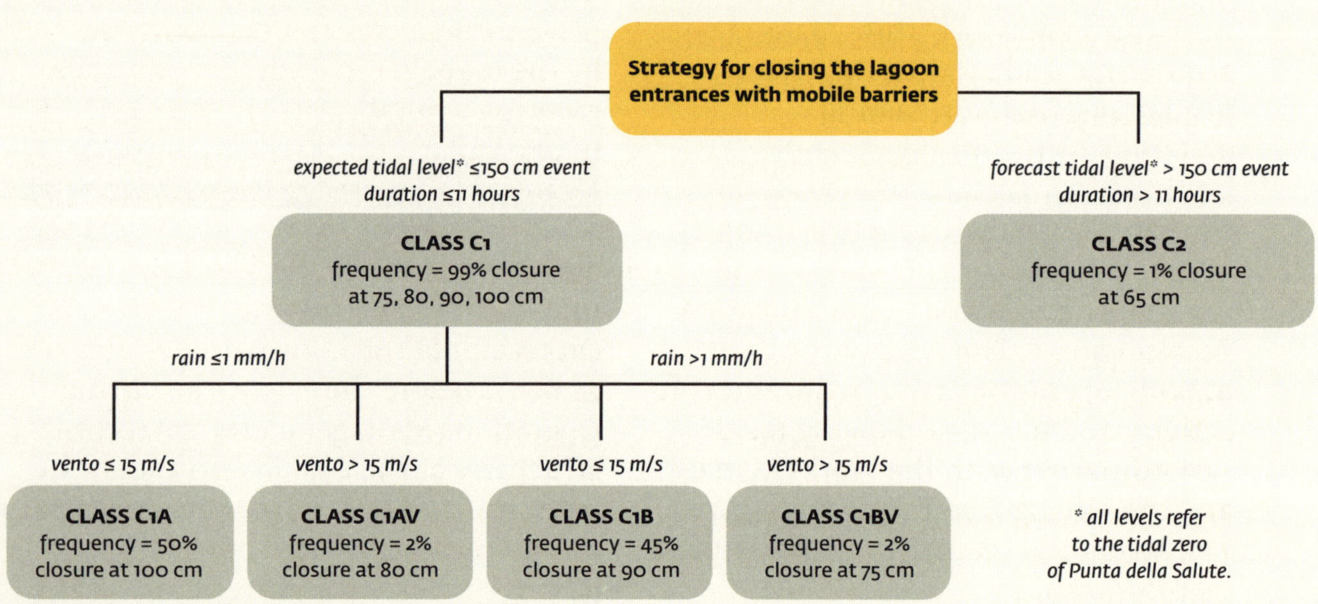
Strategy for closing the lagoon entrances with mobile barriers
expected tidal level* ≤150 cm event duration ≤11 hours
forecast tidal level* > 150 cm event duration > 11 hours
CLASS C1
frequency = 99% closure at 75, 80, 90, 100 cm
CLASS C2
frequency = 1% closure at 65 cm
rain ≤1 mm/h
rain >1 mm/h
vento ≤ 15 m/s
vento > 15 m/s
vento ≤ 15 m/s
vento > 15 m/s
CLASS C1A
frequency = 50% closure at 100 cm
CLASS C1AV
frequency = 2% closure at 80 cm
CLASS C1B
frequency = 45% closure at 90 cm
CLASS C1BV
frequency = 2% closure at 75 cm
* all levels refer to the tidal zero of Punta della Salute.

local tide forecasts, with a commitment from scientific and operational institutions. It is therefore expected that the most noticeable improvements will come over time from greater accuracy in weather forecasting, which will in any case require a development period of the order of 5–10 years, and in advancing our knowledge of energy exchange at the sea–atmosphere interface; during this period we will have to live with the current modelling uncertainty, which has, however, allowed us to manage events without any particular operational problems in the period between October 2020 and February 2021.

The operational management structures

The operational management of MOSE is assured by two horizontal technical structures: the operations manager based in the Arsenale, led by the head of the decision-making operations room; and the head of controlled operations at the lagoon entrances, led by the operations manager for the raising. Both heads employ teams of specialist technicians who work in shifts during events.

The head of the decision-making operations room makes use of: monitoring network units, data acquisition and monitoring systems display systems, forecasting systems, decision-making and issuing of movement orders systems, communication with agencies, the Arsenale infrastructure support unit, and ICT (Information and Communication Technologies). The head of the decision-making operations room ensures its full activity and has the following tasks:
• coordinating with the Works Manager and the Lifting Operations Manager;
• ensuring the coordination of the above-mentioned units;
• ensuring, when an event is detected, the composition of the shifts and that personnel have been pre-alerted;
• ensuring the coordination of maintenance activities and preparatory checks of the installations serving the decision-making operations room with operational activities;
• management of the MOSE system, with regard to the issue of lifting operations;
• applying any emergency provisions communicated to the Decision-making operations room by the competent bodies (for example, the works manager or external bodies such as the fire brigade, etc.);
• ensuring ordinary and emergency communications to the competent bodies and to the person in charge of the works;
• ensuring that the decision-making center is informed at the times provided for in the procedures.

The Lifting Operations Manager supervises operations at the various entrances and provides all necessary coordination and supervision instructions to the technical structures located at the Lido, Malamocco, and Chioggia sites. Coordination takes place through key figures, assigned by the manager in each port entrance, covering several shifts if necessary.
In particular, the Lifting Operations Manager is assisted at the various sites by a lagoon entrance manager, one for the out-of-water plant areas and one for the underwater plant areas.
On receipt of the notice of a possible lift, 24 hours before the event, the Lifting Operations Manager alerts the technical structures at the lagoon entrances, the technical and nautical services necessary for the transport of personnel, the Italian Army for the surveillance of the

telecommunications equipment between the inlets, and the e-distribution for the surveillance of the electricity supply from the national grid.
If the event is confirmed, 9 hours before the event, the Lifting Operations Manager mobilizes the entire structure of the command operations center and reaches the sites to start the activities of preparation and checking of the lifting equipment.
The lifting of the floodgates can only take place after the Lifting Operations Manager has issued a "green light" following these preliminary tests on the installations.
During the provisional operational phase, in which the MOSE system plants are not completed, commissioned, and tested, the Lifting Operations Manager coordinates with the Works Management that supervises the completion of the plants and the companies that carry out maintenance work, establishing
• the priorities of the interventions in terms of safety and the possibility of carrying out lifting operations: priority is naturally given to interventions that resolve situations that prevent lifting and/or compromise the safety of operations;
• possible corrective actions, if necessary, of a temporary nature if for technical-logistical reasons (availability of components, execution times, etc.) it is not possible to implement definitive solutions in time to guarantee a subsequent lifting operation.
This is to avoid possible inefficiencies that might prevent the lifting of the floodgates.

↑ The back up room and detail of a barrier in operation.

Some insights into operational management

In the period between October 2020 and February 2021, a series of high water events were managed in lines

with the procedures for flood defense in the provisional phase, during which decisions concerning closures were guided by a support system that took appropriate precautions conditioned by the fact that they were operating in a construction site environment, with uncompleted and untested facilities.
The MOSE Decision-making Operations Room identified the events and managed them also by undertaking dozens and dozens of off-line simulations with the two-dimensional MIKE 21 HDFM model of the Venice lagoon, installed in the IT infrastructure of the same room. Among others, simulations were carried out of possible partial closures of the barriers in order to assess in real time whether or not it was appropriate to proceed.
An ex post analysis of the events was also carried out using the same model, with reconstruction of the trend in tidal levels in the lagoon, with the assumption that MOSE had not been activated.
The head of controlled operations undertook the lifting operations at the lagoon entrances, working on the plants and their automation and control systems. In particular, he assessed the performance and effectiveness of the plant solutions adopted as the plants were completed, highlighted critical issues, and proposed improvements with the general aim of increasing the reliability of the lifting plants.
These evaluations were possible thanks to the analysis of information on the behavior of the systems, collected through data gathered over time, constantly archived by the automation systems, downstream of each barrier.
During the last autumn–winter season, the barriers at the lagoon inlets were raised on twenty occasions, testing simultaneous closures of all the barriers, early closures of the Lido and Chioggia barriers, and delayed closures of the Malamocco barrier, and closures of the Lido barriers only, depending on the weather and sea conditions expected.
During the mobilization of operational staff, it was possible to test and refine the procedures for the alert phase, also in terms of communications to and from the authorities, in particular with the Port Authorities of Venice and Chioggia, which are responsible for navigation and whose hazard warnings are essential to ensure that the barriers are closed in complete safety for all transiting vessels.
The closure of the barriers for about 44 hours on December 5 and 6, 2020, is the most interesting event of the period: it was shown that MOSE is able to manage closures that are much longer than the design limits, with 4 consecutive tidal surges and the reopening of the Malamocco barrier for just a few hours in the afternoon of December 5, 2020, to allow ships to leave the commercial port.

Waiting for MOSE. Chronicle of a historic meeting

Milva Andriolli

To have been in Venice on October 3, 2020. To have been in St. Mark's Square. To have crossed it without boots, quickly, freely. To have been there to tell Italians and the whole world that there was no *acqua alta*, or high water that day. Perhaps from that day on, for Venetians, there would be no more *acqua alta*. And that the credit was due to that yellow giant called MOSE, born inauspiciously and raised badly, hailed yet also hated, supported and rejected, mocked and ignored, growing slowly, slowly, slowly, awash in tsunami of bribes, slush funds, but also public money—and lots of it—diverted from under the eyes of Italians, under the eyes of the Venetians.
That day, however, MOSE was ready. We had to change course, to draw MOSE out of oblivion, out of the controversy and water as a single body, with pride. The message was: MOSE works.
Let's rewind a bit and look back at that day, shooting straight through the years of scandals and criminal investigations and focus instead on that day of reconciliation.
I had the good fortune to film, record, and document the emptiness, what we might call the "nonevent," the almost dry land (the paving stones were wet but only because it rained that day) and at the same time immortalize the joy, the celebration, the amazement, the incredulity, the hope of the people.
For the first time in its seventeen hundred years of history, Venice was celebrating a different kind of marriage with the sea. And its inhabitants, from that day on, had to make peace with the giant. Its seventy-eight hands, raised together, were able to shelter and protect all the islands in the lagoon.
Yes, something happened that day, something momentous and huge. A souvenir shop in St. Mark's Square. Murano glass and vases in the window. The lady is holding a bucket and

← Venice, high water on October 3, 2020

↑ Venice, high water
on October 3, 2020
← ↓ Venice, high water
on November 12, 2019

brush. She is talking to someone. They both look out, in anticipation.
I ask: “You’ve got your bucket ready, but there’s no water, so what’s going on?”
“So we’re happy. I was just saying now that I’m so happy I could go to the middle of the square and cheer. It’s so exciting.”
“So it’s party time today?”
“Absolutely.”
A little further on, there is one of the most established jewelry shops in Venice. The owner is at the entrance. She is wearing rubber boots, those that come up to below the knee. A smile lights up her face.
“For me it’s a historic date, naturally. As it is for everyone. It’s a Redentore, a Redeemer, that’s all I can say. And I can say that as a Venetian. When the high tide got to 77 centimetres, it stopped. It was supposed to go up to 130. So for me it’s a thrill and as well as a relief.”
On the opposite side of the square, among the busy waiters in the salon of the Serenissima where the “black drink” (coffee) was born, I find Signora Rita.
“Three hundred years of history for Caffè Florian in St. Mark’s Square and not a drop of water today?”
“It’s been a difficult year for everyone, as we know. And it has been for us too. The *aqua granda*, or ‘big flood,’ last year, Covid this year. This morning, I confess that we traders met in the square and we were really moved because it was such an unusual spectacle. And last night, too, we were worried. Instead, I find it a memorable, moving spectacle.”
“Venice in my opinion—and I’d like to say this calmly—is safe. This is the point. Then there’s the work, the ups and downs; the past is gone, the future will be better. But the important thing is that we don’t go under like last year. Last year ruined our lives.”
Voices of Venetians, eyewitnesses to a miracle. Voices transcribed from my first report that aired that day on national RAI news. They speak to the emotion they felt because of the result, set against a backdrop of real concerns, of unresolved fears, of nerves on edge, and of millions in damage to the state coffers. Only eleven months earlier, the *acqua alta* of November 12 had been a dramatic event: the high tide that night reached a maximum of 187 centimeters above the average. This was only 7 centimeters less than in 1966, which went down in history as the first *acqua granda*.
The phenomenon prods our memory banks. Whenever the acqua alta returns, it reminds everyone of broken promises.

On that appalling night, the question was: what happened to MOSE? That was when it struck us all. The point of no return. So you can understand that, yes, on the night of October 3, 2020, the city was expecting MOSE to charge into action for the first time. But no one had predicted what would actually happen. The image of an unflooded St. Mark's Square, with unused raised walkways, people in trainers, the Basilica looking resplendent, dry, with even its narthex, the lowest point in Venice, accessible. This is what must remain of that October 3, 2020. The good news that has done the rounds of the world: MOSE works.

It is a masterstroke. The miracle of MOSE saving the lagoon. And never mind the fact that politics dictates that the giant—still in a running-in phase and not yet fully tried and tested—only comes into operation at tidal levels above 130 centimeters. When fully operational, this will be reduced to 110.

St. Mark's Square goes underwater at much lower levels. "Normal" tides of just 80 centimeters or little more are enough to ruin the Basilica's marbles and mosaics. So, to achieve the objective and to make an impression, MOSE was raised earlier, stopping the tide within the lagoon at 70 centimeters. Numbers, so many numbers: too complicated to explain, too complicated to understand. What counts is that it works. Another thing is that the giant cannot be seen. Its four barriers at the lagoon entrances are far away. And it's not easy to link cause and effect—that is, the raising of the gates with the absence of high water.

To tell the truth, few Venetians have ever seen this giant up close, dismantled like a Lego kit or assembled and at work. Few have gone down to 20 meters below sea level, into the concrete tunnels that run parallel to it, where the hinges and gates come together. It takes considerable effort to put it all together. Human ingenuity and the force of nature. The Modulo Sperimentale Elettromeccanico (Experimental electromechanical module, or MOSE) and the tides.

Saturday morning, a few hours before high tide. Sestiere di Cannaregio, at the art gallery at the foot of the Guglie bridge. I ask the proprietor, "Are you putting it in place or removing it?" The man is holding a metal baffle used as a flood barrier to seal the entrance to shops and homes. He replies, shaking his head: "It seems there's no water. I don't understand."

A couple of lawyers walk by. One says, "If it keeps on working like this, we're fine."

I ask the other: "Do you still have any doubts?"

"Do I still have any doubts?! This is the first time, the first

trial. The first go went well. Let's hope it stays that way."
It didn't. It was a good start, but the next day, Sunday, October 4, MOSE was taking the day off. Tourists swarmed into St. Mark's Square, convinced they were capturing the last high water in Venice's history. And the residents, shrugged their shoulders: "Tourists!"
From October 3 to December4. From the scoop to the forty-hour stress test. MOSE is put under stress, day and night, to hold off three tidal peaks of over 130 centimeters. This is another key step on the road to bringing the Venetians closer to MOSE. They both have to come to terms with each other. And it is not easy. For example, having MOSE raised continuously for hours at the three lagoon entrances is stirring up controversy over its compatibility with the port. Shipping transit during this pandemic period is very limited. There have been no cruise ships for months and, all things considered, the Venetians don't mind at all. But work is scant, and merchant and fishing activity is also suffering as a result. Dockers and shipowners are on the warpath. The Chioggia seagoing community, Italy's largest, is raising the issue of complementary works such as the navigation locks, which would allow boats to pass through even when MOSE is up and running. There are two such locks in Chioggia. Poorly designed and built even more poorly, they are impracticable. More public money wasted. But when the sea cuts up rough, it's scary. The fishermen never forget it. More than that, it's forty hours straight with no change of water. The lagoon is in danger of becoming a closed pool, alarmed experts have stated, which also collects incoming water from various rivers—not an indifferent addition.
The attempt at reconciliation is effected with an experiment. Opening the Malamocco entrance to allow ships to pass, under the watchful eye of the Port Authority.
"It can be done," say the Consorzio Venezia Nuova technicians. The work is flexible. And it can be done in the dark, almost at night.
Yes, MOSE at night.
An unforgettable, memorable night between December 4 and 5. The weather outside is awful, with heavy rain and a powerful Sirocco wind blowing across the lagoon. The weather and tidal forecasts are not at all good (an unprecedented three peaks of 130 centimeters). And there's another beast on the prowl to torment the Venetians' sleep: Covid-19. It sickens, kills, starves, eats up income and capital. Like a sniper, it has struck at the heart of Venice, at its economy, its industry: tourism. The large hotels have closed. Bars, restaurants, bed and

breakfasts, museums, and shops—all closed. The vaporetti are empty. The gondolas at rest, motionless. The Fontego dei Tedeschi, which usually has three million visitors a year, is deserted.
This is Venice. Pandemic, act two. Can you imagine wearing a mask, with your gel and maybe boots, waiting for *acqua alta*?
"No *acqua alta* tonight!"
"No, no. A dream. The first night I can stay home with my wife." Paolo Ballarin, seventy-one years old, is happy. For fifty-two years he has been the biavarol (grocer) on Strada Nova. Without MOSE, how many sleepless nights has he had? Away from his wife Ornella.
A little further on. Another shopkeeper. He sells gloves.
"What are you doing with the bucket and mop in your hand?"
"Nostalgia for the *acqua alta*." He laughs.
"How did it go last night?"
"Tonight was a miracle for us. A miracle. It doesn't even seem real to us. MOSE is finally working."
At spritz time, outside the bar, I find Gino and Bepi.
"Our MOSE, after so many years. After thirty years. And all the money they spent. No talk of that. Let's hope it works out. That Venice stays dry for a bit."
"It really turned out well for us. Because with all the money they spent, it was only right that something should come of it."
MOSE supporters in Venice have never been either united or loyal. Or rather, they have been alternately so. After all, it's been so many years since the first stone was laid (May 5, 2003); the second Berlusconi government, with Giancarlo Galan as president of the Veneto Region, later arrested in 2014 and then convicted for corruption. Which we could also put in these terms: ten national governments, and four mayors of Venice. And everyone (on both the right and the left) promises—calendar in hand—that the job is almost done. The Venetians believe it... and don't believe it.
If MOSE was the mother of all kickbacks, restoring virginity to the great Unfinished Project, especially in Venice, seems neither easy nor obvious.
But its premiere on October 3 and its encore on December 4 are a blessing.
It's done. MOSE will take care of the high water. But it didn't go smoothly this time either. A shame as the Venetians had taken a liking to it. They almost didn't notice it anymore.
No more rubber boots, no more raised walkways. Goodbye to the late nights spent saving ground floor shops and houses. Goodbye to days spent cleaning with bleach, wiping away the

fishy smell of mud. Goodbye to merchandise that has to be thrown away.
Call it bad luck, if you like. In December 2020, MOSE had to deal with a series of exceptional tides: many and all close together. Something unheard of, said the experts who are worried about rising sea levels and climate change. But this is the trend, and it seems unstoppable.
As if that weren't enough, on December 8 MOSE even had Saturn against it. The weather forecasts changed abruptly. The mathematical models went haywire. Too late to call the army of engineers and electricians back to work. The control room was caught off guard. Acqua alta at 138 centimeters, and Venice went under once more.
This was a sucker punch that threatened to dampen enthusiasm and drive MOSE out of the hearts of residents. Disappointed, angry, distrustful, skeptical once again. And once again, out with the elbow grease, clean water, and bleach.
And like a mantra, the question of the 6 billion euros already spent comes up again. Of the unsustainable giant. Of unfulfilled promises. Of the narrative that is sinking rapidly.
And yet Venetians know that the mobile dam system is still in its testing phase. That the project is actually due to be officially delivered until the end of 2021. That the transition phase implies a series of tests and trials, both for the gates and for the teams that need to be trained to monitor their operation.
Has this proven to be a boomerang for those who had pushed the accelerator?
Overwhelmed by the controversy, MOSE's staff is working overtime. Three days in a row, up and down, stopping 135- and 140-centimeter tides. The giant is up to the task. It parries all the blows of the sea. It wins the challenge. Calm returns to Venice. The acqua alta season is also over. But the giant continues to work. Preparing. Training. It goes up and down.
It will have to last a hundred years.
The Venetians know this too.
Meanwhile, Gino and Bepi have almost finished their spritz.
They greet me by raising their glasses.
"Is the MOSE news good news?"
"The news about MOSE is, yes, but let's hope it lasts the distance."

→ Detail of a barrier in operation

Reproduction
Studio Pointer s.r.l., Mogliano Veneto (Treviso)

Printing
L.E.G.O. s.p.a., Vicenza
for Marsilio Editori s.p.a. in Venice